CHANCE ENCOUNTERS: PROBABILITY IN EDUCATION

Mathematics Education Library

VOLUME 12

The titles published in this series are listed at the end of this volume.

CHANCE ENCOUNTERS: PROBABILITY IN EDUCATION

Edited by

RAMESH KAPADIA
South Bank Polytechnic, London, U.K.

and

MANFRED BOROVCNIK
Institute of Mathematics, University of Klagenfurt, Austria

SPRINGER-SCIENCE+BUSINESS MEDIA, B.V.

Library of Congress Cataloging-in-Publication Data

Chance encounters : probability in education / edited by Ramesh Kapadia and Manfred Borovcnik.
p. cm. -- (Mathematics education library ; v. 12)
Includes bibliographical references and index.
ISBN 978-94-010-5563-5 ISBN 978-94-011-3532-0 (eBook)
DOI 10.1007/978-94-011-3532-0
1. Probabilities--Study and teaching. I. Kapadia, Ramesh. II. Borovcnik, Manfred, 1953- . III. Series.
QA273.2.C43 1991
519.2'071--dc20 91-34404
CIP

ISBN 978-94-010-5563-5

Printed on acid-free paper

Preface

This book has been written to fill a substantial gap in the current literature in mathematical education. Throughout the world, school mathematical curricula have incorporated probability and statistics as new topics. There have been many research papers written on specific aspects of teaching, presenting novel and unusual approaches to introducing ideas in the classroom; however, there has been no book giving an overview. Here we have decided to focus on probability, making reference to inferential statistics where appropriate; we have deliberately avoided descriptive statistics as it is a separate area and would have made ideas less coherent and the book excessively long.

A general lead has been taken from the first book in this series written by the man who, probably more than everyone else, has established mathematical education as an academic discipline. However, in his exposition of didactical phenomenology, Freudenthal does not analyze probability. Thus, in this book, we show how probability is able to organize the world of chance and idealized chance phenomena based on its development and applications. In preparing these chapters we and our co-authors have reflected on our own acquisition of probabilistic ideas, analyzed textbooks, and observed and reflected upon the learning processes involved when children and adults struggle to acquire the relevant concepts.

For this comprehensive survey, a broad range of perspectives is presented: educational, probabilistic, empirical, curriculum, classroom, computer and psychological. The intention has been to synthesize all major research in this area; a critical and analytical viewpoint has been adopted in order to re-think and re-search ideas.

As will be clear from the contents page, an international perspective has been taken by involving authors from both sides of the Atlantic. The book began to take shape several years ago at the 2nd International Conference on Teaching Statistics at Victoria in 1986; meetings have taken place several times subsequently and manuscripts shared between everyone involved. Five years is a long time but it has helped in co-ordinating the book into a coherent whole, though the individual flavour and distinctiveness of each contribution has been preserved.

We have included a historical dimension which is too often ignored; the computer perspective ensures that current ideas are also explored. We have ensured the technical soundness of the mathematics which is illustrated by a carefully chosen range of examples. Experimental work is discussed in an appropriate theoretical context and judgements on findings are made explicitly; it is only by such a critique that issues can be aired openly and constructively. Mathematical ideas are a key feature but explored within an educational framework.

The intention has been to make the book accessible to a broad audience. It is aimed at those involved in stochastics teaching at various levels. We tried to hold technicalities and jargon to the minimum level necessary to transmit our ideas and those of probability. The book gives guidelines on the discipline and related didactic research. It should help experienced teachers to find good topics for their work and also help teacher trainers to develop a deep understanding and thereby become more confident to evaluate other teaching approaches and develop their own. We particularly mark milestones for those who are involved in curricular work.

We wish to thank Alan Bishop, the managing editor of the Mathematics Education Library series for his encouragement and patient help at crucial stages. We would also like to thank an anonymous reviewer who encouraged us by describing the book as a significant addition for the international mathematical education community, yet gave advice on how to improve its readability from a textual viewpoint. Finally, we thank all the individual authors for their patience and forebearance.

Ramesh Kapadia, London

Manfred Borovcnik, Klagenfurt

Contents

Chapter 1

The Educational Perspective

R. Kapadia and M. Borovcnik

This opening chapter presents the aims and rationale of the book within an appropriate theoretical framework. Initially, we provide the reader with an orientation of what the book intends to achieve. The next section highlights some important issues in mathematical education, establishing a framework against which ideas in the book have been developed. Partly, the research has been inspired by the first book in this series on mathematical education: Freudenthal's Didactical Phenomenology of Mathematical Structures. Though he considers many topics in mathematics he excludes (perhaps surprisingly) probability. Finally, summaries of each of the chapters are related to these didactic approaches.

1. Aims and Rationale

"The Law of averages, if I have got this right, means that if six monkeys were thrown up in the air for long enough they would land on their tails about as often as they would land on their - Heads ..." (Tom Stoppard).

Chance would be a fine thing. This statement summarizes, in colloquial English language, a common view about the likelihood of a fortuitous event. It is almost as if, by being fortuitous, an event becomes less likely. There is a feeling that risks may be necessary in that fortunes favour the brave. Common language recognizes the existence of chance but it is often viewed with trepidation.

In a famous modern play, Tom Stoppard uses dice throwing to conjure up the unreality of the action he portrays. In this play, *Rosencrantz and Guildenstern are Dead* he puts at centre stage two relatively minor characters from Shakespeare's Hamlet. The scene is set by an unlikely coin-tossing sequence: Rosencrantz has just produced a 67th tail yet Guildenstern still calls heads for subsequent tosses. The odds of obtaining such a sequence with a fair coin are negligible whence the play progresses.

R. Kapadia and M. Borovcnik (eds.), Chance Encounters: Probability in Education, 1–26.

Ideas of chance are firmly established in common culture through the unpredictability of everyday occurrences, such as the state lottery or private betting or the weather. With increasing civilization, chance plays an even more important role in society. We note life insurance and quality control as specific examples, or the substantial part played by statistical methods in the acquisition and conceptualization of knowledge in various sciences. This fact is now reflected by the increasing importance of probability within school curricula in most countries all over the world.

One of the main aims of teaching mathematics generally is that it is a powerful means of communication. Mathematics offers structures for thinking. The first approaches through number and geometry were causal and deterministic. A much later but equally important approach is afforded by the "efforts to tame chance", to use Hacking's (1990) evocative phrase. Probability establishes a kind of thinking different from the logical or causal approach which is nevertheless important. The other major justification for teaching probability is that it is a significant means of applying mathematical ideas in realistic situations.

Nevertheless, probability is a difficult subject to learn and teach. Causality is much more comfortable, logical thinking is much clearer, but chance is a reality. Chance is linked to other terms like 'risk', 'accident', 'opportunity' or 'possibility'. Probability can be thought of as the mathematical approach to the quantification of chance, just as rulers measure distances. Measurement of distance is straightforward; though there are some paradoxical ideas in geometry, these occur at a rather deep level which need not influence school mathematics. In probability, paradoxes or counterintuitive ideas occur at the very heart of the subject, in the definition, and subsequently in relatively simple applications. This is borne out by the vast psychological research on people's misconceptions as well as in the difficulty experienced by pupils in applying probabilistic notions in problems. People do not seem to have probabilistic intuition in the same way as they have geometric or visual intuition.

This suggests a number of major questions requiring analysis and research. Our main focus is on probability rather than statistics. Nevertheless there are some occasions where we cannot deny the close links between the two subjects. However, we will not enter into the controversy whether probability is part of statistics or a mathematical foundation of statistics which may be separated from it. We simply had to limit the scope of our book pragmatically in order to ensure that the ideas presented are coherent. Another book should be written on topics of descriptive and inferential statistics to give guidelines for these disciplines which are, of course, also very important for school mathematics.

Even if one has to acknowledge that there is virtually no tradition in teaching probability as compared to calculus, one has to note that there are distinct approaches to probability in different cultural environments. The 'Continental European' tradition is heavily based on combinatorics and mathematics (Germany and France). In this view, statistics cannot be taught and properly understood without a properly based probability theory. The Anglo-American tradition, on the other hand, is based on an experimental approach where probability is only a possible, but not necessary aspect of an experimental approach towards statistics.

The cultural differences have led to different approaches to curricular design (Borovcnik, 1991). The influential movement of 'statistical numeracy' of exploring data and decreasing the importance of questions of probability has grown from discussion in the United States; in the Anglo-American tradition the accent is on structuring reality (how to solve real problems). In the Continental tradition the central issue is still 'stochastical thinking'; the primary accent is on structural connections (how probabilistic concepts would help to structure one's own thinking). The Anglo-Americans would take a sceptical view of how much students' thinking is likely to be structured by the imposition of formal probabilistic ideas. This is supported by the following two quotations from the current American discussion.

> "Long before their formal introduction to probability, students have dealt with countless situations involving uncertainty ... They have a coherent understanding ... in everyday situations. It is into this web of meanings that students attempt to integrate and thus make sense of their classroom experience ... My assumption is that students have intuitions about probability, and that they can't check these in at the classroom door. The success of the teacher depends on how these notions are treated in relation to those the teacher would like the student to acquire." (Konold, in press)

> "...Do not dwell on combinatorial methods for calculating probabilities in finite sample spaces. Combinatorics is a different -- and harder -- subject than probability. Students at all levels find combinatorial problems confusing and difficult. The study of combinatorics does not advance a conceptual understanding of change and yields less return than other topics in developing the ability to use probability modeling. In most cases, all but the simplest counting problems should be avoided." (Moore, 1990, p.122).

The perspectives we take in this book, are multifaceted like the topic of probability itself. The probabilistic perspective analyzes the ideas underlying the formal concepts; historical and philosophical aspects as well as paradoxes and fallacies illuminate these ideas. The chapter on empirical research reveals how individuals view certain probability concepts. The curricular perspective draws on the experiences from several countries; a special idea of how probability has to be conceived is the basis of the classroom perspective. The way that the advent of new technology will change ideas and usefully illuminate old ideas, is the scope of the chapter on computers. Finally, psychological research on probabilistic thinking is reviewed.

2. Views on Didactics

Mathematical education is relatively new as an academic field of study. There are few widely accepted paradigms, so people work from different, even conflicting perspectives. We give a brief commentary on the history of the discipline and of positions within didactics which influence our work. This could be skipped by the reader only interested in probability.

In some ways one could think of mathematical education beginning with the first deliberate instruction. The most widely known text is Euclid's Elements: as Plato wrote on the door to his academy 'let no-one ignorant of geometry enter'. There have been many texts to make geometry more accessible: the precursor of the Mathematical Association was the Association for the Improvement of Geometrical Teaching (1871). *The Mathematical Gazette*, published from 1894, was the first journal aimed at promoting good methods of mathematical teaching. However, though textbooks, associations, and journals have an important role to play in mathematical education, one cannot assert that these influences alone established mathematical education as an academic discipline.

For authors were primarily concerned with producing a textbook, they did not attempt to give reasons for their choice of topics, order of presentation, or methods of explanation. Similarly, reports of the Mathematical Association and others commented on the current situation rather than the underlying rationale. Even in the sixties, during a great period of curriculum reform, the emphasis was on changing an existing situation. It is only over the last thirty years that mathematical education has emerged as an academic discipline in its own right. This can be partly traced back to the first International Congress on Mathematical Education in 1969 and the establishment of the first international academic journal, *Educational Studies in Mathematics*. There still seems to be little agreement on the foundations of the subject though some consensus is slowly emerging.

Traditional didactics of mathematics (and of other disciplines) is concerned with the construction of curricula, selection of topics and concepts for teaching and with the design of promising methods to present topics in class. Here a broader view on didactics is taken. Didactics is devoted to relations between the 'triangle' of individual or socially organized subjects S, parts of reality R, which is the 'objective' world, and knowledge of this reality in the form of theories T (theory in the meaning of its Greek origin which amounts to a 'view on reality').

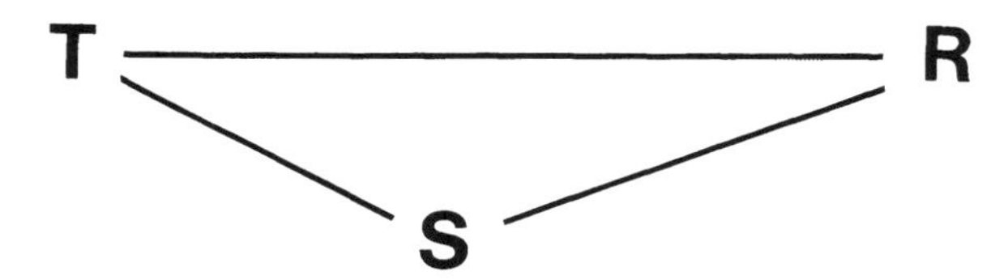

Fig.1: A relational triangle on reality R, subject S and theory T

If the focus is on the theory vertex T, then inner mathematical relations increase in importance, they yield guidance on what, how, and when topics should be taught. If the cognitive structure of the individual subject at vertex S is the starting point, then psychological analysis applies; if the learning subject at vertex S is primarily seen in a social context, then sociology and classroom observations are central. When the focus is on reality R to be modelled, then the question of how a special theory may structure reality and solve problems becomes predominant.

The analysis of each vertex in the triangle separately also sheds light on its interrelations to the other vertices. Discussion of the theoretical vertex T allows insights into how reality R should 'behave' in order to fit actual modelling, and into how subjects S may think structurally on problems, and to what extent they do not. Psychological analysis of vertex S yields insights not only into how subjects think, but also into how theories T are structured and consequently require a special kind of argument, which in turn relates the vertices T and S. It is clear that the relations between R, S and T, fully developed, are of great didactic relevance and may enhance the value of curricula construction and actual teaching as well. Important strands in the short history of didactics of mathematics as an academic discipline may be assigned to the didactic triangle.

Clearly, mathematical education is an applied rather than a pure science. It must be linked to education and its inherent values, so the underlying ideas are relative rather than absolute. There also have to be close links with classroom reality. Fundamental research from fields such as history, philosophy or psychology is important but needs to be linked to classroom practicalities. Parallels may be drawn to subjects like medicine or agriculture, where a scientific approach is essential. However, pragmatic considerations are paramount: has the patient been cured as a consequence of treatment or despite it, and are there any side effects? Is the quality and quantity of crops yielded improved by special treatment or due to weather conditions or other side effects; are changes in crop also to be expected in long term treatment? Clearly such questions cannot always be answered definitely as there are many variables which cannot be controlled as physicists manage in their experiments.

The 'New Mathematics' movement was too heavily concentrated on the vertex T; sets and logic were intended to yield a coherent structure of theory T as a self contained entity. Whereas this approach served to justify a specific formulation of theory in itself, its accessibility was beyond the reach of the learners as the relations to R and S were completely ignored. In reaction to subsequent poor experiences, the wave of integration of applications with the focus on the vertex R emerged. Inherent problems were that tackling real applications is onerous and the mathematics involved is often too trivial. Pseudo-real problems, on the other hand, induced their own difficulties because of their artificiality. The psychological strand draws heavily on the vertex S in investigating subjects' thinking; it has to cope with problems around eliciting and generalizing strategies and linking the findings to a coherent theory. With the advent of computers an old position in didactics has been revived: problem solving. This is a mixture of cognitive strategies on the S side, 'real' problems on the R vertex, and logical thinking on T, all intended to disclose T to the learner.

We describe four different modern positions in the didactics of mathematics with respect to the didactic triangle. Fischer's (1984) views on 'open mathematics' are developed to enliven 'closed mathematics' for the learner. This process oriented position between T and R integrates the learner S. Fischbein (1987) relates T and S via a dynamic interplay of intuitions and mathematics. Freudenthal's (1983) didactical phenomenology makes phenomena in R the base for structural, theoretical deliberations. Bauersfeld's (1985) delineation of subjective domains of experience make the relations between S and R via personal socially mediated experiences a starting point.

Fischer's open mathematics

Learning mathematics is not a straightforward task; this should be no surprise when one remembers the struggle and searching involved in discovering the ideas initially. Fischer's open mathematics is a means of helping to bridge the gap to formalized mathematics in the learning process. A mathematical theory establishes relations between 'objects'. These relations indirectly determine notions and concepts of the theory. They are codified in sequences of symbols and rules for conclusions. This is how progress in mathematics is made and how relations are justified and communicated. Within formal or official mathematics, however, only a small proportion of the many relevant relations are retained.

The mathematician who derives a new theorem has a broader spectrum of ideas in mind which are not covered by the symbolic proof. These ideas are, for a time, shared by the scientific community and they are essential for the new results to be understood and acknowledged. But only the relations connecting terms within a mathematically (axiomatically) based theory are subsumed under the heading 'closed mathematics'. This means, that T is (ideally) completely separated from R and S. Closed mathematics is thus self contained and static. Many relations relevant for understanding notions are excluded by this approach.

The situation poses a dilemma for teachers in facilitating the learning process. The spectrum of relevant relations is vague and often is lost when a closed version is established; but these relations of codified mathematics are not understandable if one does not regard the broader but vague general relations between notions and concepts. This marks a complementarity between objective relations covered by the codes of mathematics and the subjective relations in our mind, both being inextricably interwoven. Open mathematics is designed to solve this dilemma. The trick is to go back to the process of mathematics in *statu nascendi*. In the process of evolving new mathematical terms the subject S is put into the centre of the stage. In model building one has to respect reality R and to use and develop mathematical pre-concepts and concepts from T in order to 'communicate' with the problem.

Within the usual conception of closed mathematics, the intention is to develop general theories. If specific theorems are unprovable, the mathematical model is supplemented by additional assumptions without paying attention to any 'reality'. General and theoretical criteria are able to render an optimal solution for a theoretical problem. Fischer (1984, p. 140) comments on the intention of closed mathematics:

> "If a field is structured mathematically, then one trusts in having covered *all cases*, in being able to answer any related question, that a *closed theory* is available. ... This drive for closed logically consistent structuring is a characteristic feature of mathematics and a necessity for its utility."

One can try to open up closed models through feedback and discussion of underlying ideas. Open models are only effective in connection with the environment, they are in a dual interaction with this environment, separated from it they establish no solution. Moreover, models only mathematize a part of reality, it may even be the case that there are rival models yielding different solutions with no clear means of choosing which is best. Open models do not yield a unique solution but a means of structuring reality.

> "[There is another aspect of mathematics,] an aspect which may be characterized as being *preliminary*, which ... deals with *approaches to the formation of concepts*. The ingredients which are developed in open mathematics are, above all, far reaching concepts with a high degree of self explicacy. ... Those concepts increase in importance which are very easy understandable and may thus be useful for communicating." (Fischer, 1984, p. 140-141).

Open mathematics, in contrast to closed mathematics yields possibilities. The various partial modellings are only different representations of the problem. Mathematics regains its character as a means of orientation, of structuring reality. As the communication between those involved increases in importance, two components of terms and concepts become highly relevant; concepts which are easy to understand and directly communicable, not requiring jargon nor a whole theory; and terms which may be visualized and are therefore easier to grasp. Open mathematics serves as a device to revive closed mathematics by changing it again into its emerging state, the learner being the (guided) researcher. It may also be an opportunity to initiate more sensible applications of mathematics.

Fischbein's interplay between intuitions and mathematics

We have been referring to notions and concepts as being established as relations between objects. The learner S and interrelations to a codified theory T are included in psychologically oriented approaches to didactics. Fischbein (1987) develops a dynamic view of the relations between intuitions on the S vertex and mathematics at the T vertex, the impetus may come from reality as a problem per se or as a task set by the teacher. He distinguishes primary and secondary intuitions:

- Primary intuitions can either be a motivating power or an obstacle for the (historical) development of a theory or for an individual's reconstruction of concepts, i.e. the developing comprehension of the learner.
- Secondary intuitions are what should emerge from the theoretical penetration of a concept field or from a systematic teaching of this concept field.

A dynamic process of interplay between the different levels emerges during the phase of development of new mathematical terms as well as in the process of learning mathematics. Thus theory as well as intuitions are gradually emerging and changing. Proceeding from raw, vague, primary intuitions I_0, a model M_0 emerges as part of a preliminary mathematical theory. This is accompanied by a change at the level of intuitions: neither the mathematical model M_0 nor the secondary intuitions I_1 resulting from M_0 coincide with the primary intuitions I_0. The intuitive conceptions I_1, on the other hand, are a starting point for further attempts of mathematization and so on. This is of minor importance for codified mathematics; but the interplay is vital with regards to learning mathematics.

The terms 'intuitive conceptions' and 'mathematical model' as a specific part of a mathematical theory are vague. We do not define them separately but explain them only by their mutual interrelations. Intuitions represent immediate knowledge, a net of relations which directs comprehension and action. This net of relations includes belief systems and the cognitive structures associated with a certain mathematical model. A mathematical model, on the other hand, is an objective and hence communicable representation of intuitive conceptions.

Fischbein ascribes various features to the notion of intuitions. Intuitions are self evident, they allow extrapolations of conclusions, they are global and synthetical. Intuitions exert a coercive effect on the process of conjecturing, explaining and interpreting. The dynamic view of Fischbein has already been alluded to above. This approach to explain the notions of intuitive conceptions and of a mathematical model depends on a kind of complementarity in the sense of Otte, i.e. one notion cannot be explained without reference to the other (cf. Bromme et al., 1981, 205-227). Furthermore, intuitions do bear a strong visual component.

The interplay between intuitions and mathematics has been used by various authors to reconstruct the historical development of mathematics as well as to describe, explain and understand the cognitive development of an individual. Fischbein has developed pertinent ideas to describe difficulties in learning as has Wittmann (1981). Malle (1983) gave a reconstruction of the historical development of the concept of number by means of revealing this interplay.

A different approach for the didactic reconstruction of concepts is founded on the complementarity between actions and reflections: Piaget (1951), Dawydow (1977), Malle (1983), Dörfler (1984) and others. These approaches develop a dynamic between the *R* and the *S* vertex of the didactic triangle, thereby (re)establishing parts of the theory. Of course it is possible to construct links to the former complementarity of intuitions and mathematics. Intuitions accompany actions and also emerge from actions. These intuitions enable or hinder the development of a theory referring to the action at the outset.

Freudenthal's didactical phenomenology

Freudenthal establishes his approach towards didactics of mathematics at two levels. The first level concerns the relations between reality and theory.

> "Mathematical concepts, structures, and ideas serve to organize phenomena from the concrete world a well as from mathematics ... By means of geometrical figures like triangle, parallelogram, rhombus, or square, one succeeds in organizing the world of contour phenomena; numbers organize the phenomenon of quantity. On a higher level the phenomenon of geometrical figure is organized by means of geometrical constructions and proofs, the phenomenon 'number' is organized by means of the decimal system. So it goes on in mathematics up to the highest levels: continuing abstraction brings similar looking mathematical phenomena under one concept - group, field, topological space, deduction, induction, and so on." (Freudenthal, 1983, p.28).

Phenomenology of a mathematical concept means describing this mental object (nooumenon) in relation to the phenomenon it is describing; this reflects the interplay between theory and reality. Mathematical concepts are thought objects which may partially be explained by the phenomena which are structured by the former. Numbers are thought objects but working with them can be a phenomenon shedding light on this thought object. Even at this first level of phenomenology, the individual subject *S* of our triangle plays an important though implicit role. *S* is intended to acquire the mental objects underlying the mathematical concepts which organize phenomena he/she experiences.

Like Fischer, Freudenthal abandons the logic of closed mathematics with *T* completely separated from *S* and *R* as it turns the original character of concepts upside down; his focus is on how the relations of an existing theory *T* could supply structure for real phenomena from *R*. Common to both approaches is that the model building subject *S* needs an intuitive access to the theory level *T* to see how mathematics successfully models *R*; this is achieved via easier concepts so that mental objects can be developed. These mental objects largely coincide with Fischbein's intuitions. Phenomenology is an interplay between reality as a starting point and theoretical concepts in which the subject has to acquire adequate mental objects to see how reality is structured. It is precisely these mental objects which allow the subject to understand the formation of theoretical concepts, they establish the intuitive pre-concepts upon which theoretical concepts may be built.

Phenomenology is within the scope of didactics but Freudenthal's second level of approach is didactical phenomenology.

> "If in this relation of *nooumenon* and *phainomenon* I stress the didactical element, that is, if I pay attention to how the relation is acquired in a learning/teaching process, I speak of *didactical* phenomenology of this *nooumenon*." (Freudenthal, 1983, p.28).

Didactical phenomenology shows the teacher where the learner might step into the learning process of structuring phenomena by theoretical concepts. Phenomenology of describing how a mathematical structure organizes reality may be seen as a cognitive product, the didactical counterpart as a cognitive process. The basic ingredients are phenomenology proper and insights in the learning processes. The cognitive process of mental growth might be another source but Freudenthal holds the converse view that this process may not be understood without didactical phenomenology.

> "Where did I look for the material required for my didactical phenomenology of mathematical structures? ... I have profited from my knowledge of mathematics, its applications, and its history. I know how mathematical ideas have come or could have come into being. From an analysis of textbooks I know how didacticians judge they can support the development of such ideas in the mind of the learners. Finally, by observing learning processes I have succeeded in understanding a bit about the actual processes of the constitution of mathematical structures and the attainment of mathematical concepts." (Freudenthal, 1983, p. 26).

Thus, Freudenthal believes that one should search for didactically fertile problems which are neither too simple, nor too special, nor too sharply formulated. It is such problems that compel the invention of concepts. For human learning is pre-programmed in such a way that a few examples will suffice; but it takes much less trouble to sprinkle a learner with a shower of examples than to search for the one that really matters. For mathematical concepts arise out of a specific context to meet a particular need. After a concept has been invented, its domain of application is extended and the process of generalization removes the concept from its original setting.

Probability is the theory which organizes the world of chance phenomena. One has to start with phenomena which beg to be organized: intuitions about unpredictable events, games of chance, occurrences which seem to happen without regularity etc. One then has to teach the learner to constitute the mental object for himself/herself. This is the stage preceding concept attainment but too often ignored. Freudenthal believes in applications coming before concepts.

His distinction between concept attainment and constitution of mental objects illuminates his approach. In everyday matters we need not learn e.g. the concept of a chair, we learn by concrete objects, what a chair is. This is the same with simple mathematical objects like circles. However, mathematical concepts are susceptible to more precision. Therefore, one might be tempted to teach rigid concepts, made accessible didactically by an embodiment which, however, unfortunately often represents a false concretization. This kind of concept attainment of teaching abstract concepts by concretizing them is judged to be an antididactic inversion by Freudenthal.

Bauersfeld's subjective domains of experience

For Bauersfeld, a theory of didactics of mathematics has still to be formulated. He distinguishes two aspects of didactics: the instructional practice of transmitting knowledge and the social practice of interactive constitution of that knowledge. Indeed, one of his central points is that knowledge is not immutable but changes even its character in the teaching/learning process. The first variant dominates common teaching practice and is best characterized by viewing theory to be learned as fixed. Accordingly, didactic efforts focus on the structure of the discipline as the supreme guide line, or, on a hierarchy for learning. This leads to a bias either towards mathematics or towards psychology. Radical constructivists hold that the meaning of mathematical concepts is established by social negotiation, solely by interaction between subjects. They deny that there are some objective structures of knowledge at vertex T without an interaction between several subjects S. The second variant of mathematics teaching delineated above refers to the construction of meaning by interaction and not by reconstruction of stable theoretical truths.

Bauersfeld's ideas are a weak version of this sociological paradigm. He stresses that subjects S are not isolated from one another in the didactic triangle but are socially organized and interacting. Learners actively construct meaning, whereby social interaction and their experiences are basic influences. This subjective experience is organized within a 'society of mind' about which Bauersfeld (1985, 11-17) develops several theses.

1. Each subjective experience is bound to a specific context, i.e. the experiences of a subject consist of subjective domains of experience.
2. The whole body of subjective experience presents itself in an accumulation of non hierarchical subjective domains of experience (SDE) the 'society of mind' ... - which compete for being activated ...
3. The decisive basis for the formation of an SDE are the actions of a subject and the network of meaning construed by him/her, or, to be more precise, their representation in the social interaction.
4. There are no general concepts, strategies, or procedures. The subject is capable of thinking about them in an abstract form, but they are not at his/her disposal in this abstract form, i.e. they are not capable of being activated free of context.

Learning is global with all the senses involved. A subjective domain of experience is more easily activated if it is loaded emotionally and if it has been activated frequently. The structure of this society of mind is non hierarchical at first sight. The isolation of the various domains of experience is a barrier for revising and refining understanding in the learning process. A 'more general' concept will be retrieved in a distinct domain according to the circumstances of its genesis. It is difficult to recognize that this is the same as some lower level understanding of the concept, so that the new experience should be stored in the same domain of experience as this requires the comparison of two different domains. As use of language is also bound to context, further complications arise. The meaning actually construed to domains of experience by social interaction is influenced and facilitated by means of representation. The growth of this society of mind leads to an accumulation of domains and to a growing network of interrelations between them. On the one side, the more general domains combine perspectives of other domains and form a hierarchical structure; on the other hand, these subdomains may still directly conflict when activated. The network is difficult, as Bauersfeld's fourth thesis expresses.

3. Basic Ideas of the Chapters

This section shows how the above ideas have guided the various contributions to this book. It provides an orientation so that the reader may know where specific issues or aspects are discussed. Links and differences between the chapters will become clearer.

A probabilistic perspective

This chapter deals with the historical and philosophical framework, the mathematical background and paradoxes. The key to this chapter is its organization around ideas: how ideas developed historically, the philosophical connotations they bear, the mathematical development and, how paradoxical situations arose from conflicting perceptions.

These ideas exemplify crucial parts of Fischbein's interplay between intuitions and early formalization by theory. Within the perspective of open mathematics these ideas are fundamental keys to decode closed mathematics, as they refer to the broader relations which the mathematicians had in mind when they developed concepts to solve specific problems. In Freudenthal's terms these ideas represent his mental objects which precede abstract concept formation; their many facets shed light on possible subjective domains of experience.

Probability concepts are basically indirect and theoretical. It is hard to see why these concepts successfully structure reality. This might account for the tardy emergence of probability. Early history is marked by the rivalry between the symmetry idea which led to the rule of favourable to possible, and the economic value of an uncertain situation which led to the concept of expectation. Theoretical progress made via the law of large numbers reconciled these and paved the way for a new interpretation of probability as relative frequencies. The impetus from error theory and the new differential calculus led to further advances. The central limit theorem provided a basis for the normal distribution of errors and justified the mean as the best value from a series of repeated measurements.

It is paradoxical that the theoretical progress via the law of large numbers and the central limit theorem was neither accompanied by a sound mathematical foundation nor a philosophical clarification of the probability concept. The first definition by Laplace was a conceptual drawback compared to standard theory. Ambiguity around the 'principle of insufficient reason' diminished the prestige of probability in the early 19th century.

Attempts to give a sound axiomatic foundation based on a specific interpretation of probability such as relative frequencies failed. It was not before the scientific community gave up the endeavour to clarify directly what probability is, that an axiomatization was successful (in 1933). Kolmogorov's axioms *structurally* analyze the defining relations for probability: it is only the definitions and the whole body of theorems derived from the axioms which indirectly clarify what probability is.

Thus philosophy still strongly influences the interpretation of theory in probability. Philosophy is basically devoted to analyze how things are, and how a certain idea may be clarified by developing a theory around it, or, how competing ideas are represented within a specific theory. The main philosophical approaches are analyzed in detail, at least as they relate to school mathematics. The classical approach of equal likelihood is the most common starting point. The frequentist view provides an important interpretation for applications. The subjectivist view is intuitively attractive, while the structural approach seems to be free of complications only superficially.

The mathematical section is far from standard because of the endeavour to locate key ideas. Besides the introduction of the usual concepts and the presentation of two important theorems (law of large numbers and central limit theorem), probabilistic situations are outlined for two purposes. They show that real problems may be structured by a few standard situations, and that stochastic concepts may be drawn from a few distributions since the inherent assumptions establish what stochastic thinking might be.

Paradoxes and puzzles always mark a conflict in ideas; they clarify conceptual difficulties, a good reason to discuss them in class. The presentation is grouped in four categories: chance assignment, expectation, independence and dependence, and logical curiosities. The first three categories deal with specific concepts of probability, while the last deals with the distinctiveness of probabilistic thinking from logical or causal thought.

Causal thought is deep-seated and often capable of being activated against the probabilistic perception of a situation in the sense of Bauersfeld's domains of experience. The overlap with logical thinking is a paradox in itself. The formation of the probability concept within a mathematically and logically organized theory does not mean that probabilistic reasoning itself (as a theoretical type of reasoning), follows the structure of logic. In particular, the transitivity of probabilistic conclusions is not warranted.

Empirical research in understanding probability

This chapter studies the interplay between intuitions and mathematics, presenting a reinterpretation of the empirical research of a subject's understanding of probabilistic concepts.

Freudenthal urges empirical researchers to base their work on his didactical phenomenology and to deal with the full complexity of concepts, to analyze items from reality for their phenomenology, and to see if subjects themselves have mental objects available to structure concepts.

Open mathematics is also a good reference for interpretation, as the stress is on communication, either as communication of a single subject or as communication between several persons dealing with the same problem. There are promising links between the 'creation' of mathematics, the reconstruction of concepts in learning, and the process of communication in related interviews with subjects. The barriers for successful communication are the same in each case.

Bauersfeld's subjective domains of experience suggest that the reason for subjects' diverging solutions is often not a deficiency in specific concepts but in their subjectively reconstructing items according to signalized experience which might even be more appropriate. This suggests searching for possible domains of experience in subjects' answers and noting whether the context of items activates these domains.

The interplay between intuitions and mathematics, however, was chosen, as this best allows the exploitation of the dynamics of communication in an interview. It seems more generalizable to refer to somewhat stable patterns of intuitions than to possibly highly idiosyncratic domains of experience. The dynamics of vague intuitions which are not communicable without a partial theory, which in turn revise primary intuitions, may shed light on the difficulties of eliciting these intuitions from a particular item and the subjects' responses.

Researchers aim to investigate the vague pre-conceptions subjects hold. To initiate the communication, the subject has to be confronted with images and problems related to theory. Any intervention on the interviewer's side may be interpreted as partial instruction which might alter the intuitive understanding of the subject which is the focus of research interest. Even the item and its layout may be regarded as partial theoretical impetus which starts to change subjects' intuitions. If the interplay between an interviewer's theoretical demands and a subject's intuitions breaks down, communication may become meaningless.

The research experiments are discussed according to various problem fields and concepts: sample space and symmetry view; frequentist interpretation; independence and dependence; statistical inference. These categories largely coincide with the section on paradoxes of the previous chapter. This coincidence should not be surprising as paradoxes are historical items which highlight conceptual difficulties. The problem fields are discussed in the light of the background interplay of intuitions and mathematics.

The aim was to develop a scenario of various possible approaches to each item. This multiple reconstruction of items is related to the interplay between intuitions and mathematics. Many of the subjects' solutions which seemed to be wrong from the official mathematics perspective are shown to be sensible solutions; on the other hand false preconceptions might lead to correct answers. Thus interpretation of subjects' answers is by no means a simple task.

Specific categories of problems which might obstruct sensible communication are described. Extrinsic layout of the interview which ought to have no influence on subjects' behaviour, sometimes does; the items can be artificial or pseudo-real; each type of item causes its own problems; 'failure' may be simply caused by the overriding common sense of the subjects. The gap between concepts at the theory level and problem solving strategies which are much more accessible to intuitions might also obstruct sensible communication.

The discussion yields deeper insight into research findings, into the complexity of items and the world of intuitions. The results have direct impact on teaching, as communication in class might break down for the same reasons; hints for teaching are provided. Approaches which respect this interplay of intuitions and mathematics have yet to be developed. Integration of applications might help but has its own deficiencies, and, after all, may not disclose the peculiarity of probabilistic thinking.

Analysis of the probability curriculum

Each of the didactic positions discussed make clear that a planned curriculum may not be directly executable. There are many stages in the process where learners might be led astray. Furthermore, the sources of curricula may not lie exclusively in the discipline itself.

The discipline as a closed theory is organized by logical criteria which do not cover the broad spectrum of relations which were active in the state of emerging. Open mathematics suggests enlivening parts of the theory again; this involves giving direct access to concepts; to show that they must serve a certain purpose for which they have been designed; or to reveal that they are bound to a restricted core of assumptions. This would at least lead to a rearranging of concepts if not to a shift towards a 'new' theory with simpler concepts which permit easier communication.

Freudenthal states that the organization of real phenomena is turned upside down within closed theory. He advocates an analysis of how concepts could organize real phenomena and how pre-concepts in the form of his mental objects could be established. This would lead to a greater stress on applications and a rearranging of content according to its modelling capacity. It would also serve as a way to discriminate between development of mathematical concepts and the question of justifying these concepts by a logical analysis.

Fischbein's interplay as well as Freudenthal's mental objects stress the importance of focusing on ideas which are intuitively accessible so that the steps of formalization become sensible, and do not restrict teaching to logical connections alone. This would have the consequence of integrating means of representation which are easier to understand, and of integrating discussion of possible misconceptions of theory. The former may be done by exploiting the analogy of the teaching situation to the interview in empirical research, the latter by the discussion of paradoxes in class.

Bauersfeld's subjective domains of experience stress the importance of enabling learners to actively revise their domains of experience which is similar to the progressing development of intuitions via Fischbein's interplay. It would lead to striving for learning as a uniform entity, with all senses engaged and with an active role for the learner to facilitate the conditions for structuring the hierarchy of domains of experience. It would dismiss all ambitious, discipline oriented approaches to the curriculum as being doomed to fail.

These conditions are explored in this chapter by focusing on the triad of 'ideas - skills - inclinations'. This triad might serve as a key to all approaches of didactics above. They are by no means drawn from a logical body of the discipline, they refer to the world of intuitions as well as pre-concepts, or mental objects in Freudenthal's sense. They also allow students to revise their subjective domains of experience. For instance, comparing probability statements to the causal approach lies at the heart of understanding probability. Causal intuitions are very deep-seated and hinder the acquisition of probabilistic concepts, decreasing the inclination of learners to apply what they learn outside school. Many other ideas are described which should and could structure probabilistic thinking.

There are many stages of a curriculum; a distinction is made between five categories: planned - taught - learned - retained - exercised curriculum. Each fixes one crucial point where the message might fail to arrive. Exercised curriculum consists of those ideas and skills students are actually inclined to call on after the course, in their personal lives or in subsequent schooling. Traditional curriculum, however, centres its efforts on the planned curriculum. Curricular debate would be improved by revealing its sources. The traditional source lies in the academic discipline; needs of society and psychology fit with other didactic approaches. However, tradition (what was taught yesterday) and teacher knowledge are the most important influences. Student readiness is a crucial point in curricular construction, but is too often ignored. Basic categories are outlined: students' interests, their mathematical pre-requisites, the link to their own, possibly naive ideas, and how to foster transfer of ideas outside the classroom. There are a host of ideas to be drawn out.

Curricular efforts are described in detail and classified. Ad hoc activities lack coherence, supplementary units are more comprehensive than a haphazard collection of activities, but there is a lack of continuity with respect to the rest of mathematics. Some mathematics schemes have incorporated probability as part of the overall syllabus while others have a separate stochastics course at secondary level. Finally, criteria to judge the effectiveness of a curriculum are presented and research needs identified. It is interesting that this contribution does not offer a planned curriculum; it offers an overview of the questions and issues.

The theoretical nature of probability in the classroom

All the didactic approaches discussed deny a predominance of logical structure of the discipline as the ultimate guide line for education. Instead, they all establish specific types of relations which connect concepts but are not contained within a logically ordered closed theory.

Open mathematics makes clear that sensible communication on a problem should be a starting point for teaching. Concepts cannot be defined sharply from the onset, they are revised or even rejected during the process of emerging mathematics. Fischbein's interplay indicates that the teacher has to pay full attention to the intuitive level of subjects, in order to develop their intuitions. Bauersfeld's subjective domains of experience likewise suggest the need for a suitable context of experiments and for feedback on subjective notions. Freudenthal wonders how to develop applications rich enough to reveal the organizing potency of mathematical concepts and yet tractable enough to let learners really develop their mental objects and see how mathematics structures reality.

The key idea of this chapter is to characterize school mathematical knowledge as being of a theoretical nature. Meaning of concepts, according to this view, cannot be completely defined by an axiomatic foundation itself. Meaning is only covered by the fully developed theory. This theory is developed by applying available concepts to real problems. Neither the basic definitions and rules nor exemplary applications account for the concept of probability; the dual interplay is crucial.

Theoretical concepts are the link between objects of reality to be modelled and the mathematical signs used to model and describe this object. This idea is established via a relational triangle of object, sign and concept. Probability concepts and their meaning depend on the level of theory developed and on their representations and means of working with them. This complementarity between theory and application and the emerging theoretical character of knowledge both have a direct impact on teaching in class. It is, thus, not possible to define basic concepts precisely at the beginning of teaching and then to enlarge knowledge by merely accumulating new theorems and applications in a piecemeal way without changing the meaning of previous concepts.

Traditional teaching is governed by the idea of mathematics being internally consistent, a self contained closed theory. The basic assumption seems to be that mathematical knowledge can be acquired in a linear way. The axiomatic foundation contains, in principle, all this knowledge, it has only to be (additively) presented. Two types of teaching emerge: the naive concrete and the logical approach; neither leave space for a process of developing the complementarity between objects and changing concepts. The concrete approach starts with experiments in ideal games and symmetries to define probability; the internal consistency approach is based on Laplacean probability. Both miss the full meaning of the concept. Probability is neither an empirical nor a mathematical property, but the combination of these aspects bound to a context. This tension between empirical and theoretical aspects and the non reducibility to one of them is illustratively shown by an analysis of Bernoulli's law of large numbers. The same complementarity is vital for other crucial probabilistic concepts.

The relational triangle of object, sign and concept suggests a differentiation between the forms representing knowledge (the level of signs), activities or applications of this knowledge (the level of objects), and finally knowledge itself (the level of concepts). The means of representation and activity allow the development of concepts and the formation of new knowledge. The complementary character of the means of representation of knowledge is outlined for the distribution concept and tree diagrams. These types of representation are, at a higher level, again interrelated by allocating them to empirical and modelling aspects, and by characterizing them as being dual concepts.

Teaching processes aim to facilitate the acquisition of knowledge, so they have to organize development of means of representation and of opportunities for students to become active. Theoretical knowledge is not directly accessible. The logical ordering does not represent the whole explanatory potency of stochastics. The organization of theoretical, systematic elements with empirical ones is judged to be best met by a system of interrelated tasks. Problems in developing task systems are discussed and their potency is shown by an exemplary task.

Computers in probability education

Probability concepts and their meaning depend not only on the level of theory and on their representation but also the means of working with them. Tools to represent knowledge or to deal with knowledge have a sizeable impact on subjects' individual formation of this knowledge. Clearly, the advent of computers mark a huge change.

In the open mathematics approach, concepts should become easier to understand by facilitating the necessary communication. This may be via graphical or dynamic representation of models which may yield empirical meaning for the concepts involved. Fischbein's interplay of intuitions and mathematical concepts may be obscured in difficult relations like the central limit theorem. There are various means to enable the learners to enter the dynamics of intuitions and formalization. Theoretical concepts can be made experiential by physical simulation, by analogies or by visualization. Freudenthal's didactical phenomenology analyzes complex real problems in order to facilitate insight into how mathematics structures reality. The formation of mental objects as pre-concepts is crucial but may be obscured by tedious mathematical calculations. A full appreciation of the modelling potency of mathematics is not possible without varying model assumptions and re-analyzing the problem. Bauersfeld's subjective domains of experience are sensitive to representation of knowledge and the media used for that purpose. The means whereby the learner engages with problems strongly influence emerging domains of experience and their interrelations.

Within all the approaches the computer will radically change available means of working on problems and means of representation. Concrete graphical representation of models make them easier to grasp. The task of making theoretical objects experiential may be enormously eased. Complex real problems will become tractable, multiple re-analysis by varying the model assumptions becomes more manageable. The changes and their consequence are the themes of the computer perspective.

Certainly, computers have revolutionized statistical practices of data analysis and simulation. A new type of rationality is based on proof by simulation. Algorithms in iterative or recursive form yield quite a different representation of problems than the closed models and closed formulae they are beginning to replace. A change of assumptions and re-analysis is much easier than in closed analytical models. Visualization and graphical methods facilitate representation of models at different levels of abstraction offering the possibility of interaction.

The new means computers offer for representing and exploring relationships are illustrated in three typical examples; the birthday problem, medical diagnosis supported by Bayes' formula, and binomial probabilities. Sensitivity analyses of values of parameters and of assumptions are easily done by computers. Instead of a closed formula, a description by a simple recursive system function may suffice; this can be solved iteratively, or by a simple enumeration of (equally probable) cases with the computer. The programming involved is not a mere technical device but a tool for thinking. This may lead to a restructuring of mathematics from an algorithmic point of view.

Simulation as a professional tool competes with analytical methods to solve problems. The method supports the algorithmic view for modelling processes. It simplifies mathematical technicalities to derive a solution but it does not provide a black box which may be applied blindly. In order to use the technique, the underlying assumptions have to be brought out before the simulation is done. Among the advantages of simulation by computer as compared to physical random generators, are possibilities for a large number of repetitions and extensive exploration of assumptions. The use of computer generated random numbers, however, marks a crucial point. How is it possible that randomness is generated by a mathematical algorithm?

The computer perspective establishes an overview of existing software while developing a theoretical framework for pertinent research. A package which meets the theoretical and didactic requirements is still not available. There are two promising strands of development: developing special packages to fit educational needs; or using existing professional software in statistical analysis and interactive statistical languages.

Psychological research in probabilistic understanding

The earlier chapter on empirical research reinterprets work on subjects' thinking in probabilistic situations by relating and contrasting findings to mathematical concepts. This chapter is also devoted to subjects' thinking processes but with the intention of establishing genuine psychological theories on how subjects process information and the general strategies they use. For the various paradigms, there is a description of the tasks used, subject behaviour and theories of subject behaviour.

Open mathematics, with its stress on sensible communication rejects supposedly unique normative situations and puts more weight on the relations between a task and the individual subject. Feedback on how a subject perceives a task is decisive not only for developing individual concepts but also for investigating how people think.

Fischbein's interplay directly refers to subjects' internal processes when solving problems: any extrinsic action is driven by an intrinsic idea; this is his dynamic theoretical explanation of why people apply a specific item reconstruction and strategy to solve a problem. Fischbein's approach would also suggest that no empirical evaluation of a theory of cognitive behaviour be made without the necessary feedback on subjects' internal images on the item.

Freudenthal's mental objects which emerge from the organization of real phenomena, require the use of realistic and rich examples as items; no feedback is possible with artificial items, as artificial items do not serve to organize real phenomena.

Bauersfeld's subjective domains of experience suggest that any cognitive behaviour is highly context bound so that it might be misleading to search for general strategies and theories for application.

Contrary to the suggestions of these didactic positions, the psychological research all too often concentrates on an experimental situation and theoretical position which are both too narrowly conceived to explain cognitive behaviour. Many paradigms are reviewed in detail: probability learning, Bayesian revision, disjunctive and conjunctive probabilities, and correlation. The traditional approach investigates whether cognitive behaviour follows a normative pattern, using physical stimuli and word problems, which are thought to be uniquely solvable. Subjects' strategies are compared to theories which are formulated according to mathematical considerations.

The judgemental heuristics approach partly relates to Fischbein's intuitions even though no dynamic development is assumed. It analyzes the organization of behaviour in random situations by means of a few intuitively based heuristics; but no attempt is made to explain when and why a specific heuristic is applied.

The structure and process models of thinking is an approach which aims to build upon the didactic positions. Accordingly, individual information processing is guided by intuitive and analytic modes of thought. A central processor reacts to the framing of items from outside and activates sensory system, working memory and long term stores which encompass the knowledge base, the heuristic structure, the goal system, and the evaluative structure. The complex system not only includes heuristics but may also explain their application. The modes of analytic and intuitive thought are responsible for the concrete item reconstruction by the subject which in turn determines the strategy applied and the solutions reached.

Probability calibration and event-related brain potential research are recent approaches to study subjects' responses to randomness. Subjective surprise is quantified by a peak in the EEG-electrodes which measures the brain potential.

A separate section of this chapter is devoted to critical dimensions of educational relevance, focusing on the conception of the task, of the subject, and of the task-subject relation. The strengths and limitations of the various approaches are highlighted. The following conceptions of man are discussed: rational animal, irrational animal, bounded rationality, collectively rational. It is noted that psychological research often neglects the conceptions of socially organized rational subjects.

Finally, developmental approaches on the acquisition of the probability concept are reviewed. Piaget and Inhelder believe that cognitive development progresses in stages and that probabilistic understanding is bound to the formal level of abstract operations. This does not explain why probabilistic pre-concepts are found in very young children; Fischbein's approach could be the key. Information processing is also posited as a radical alternative; capacity of information processing is a crucial variable so subjects may have adequate concepts, but still become overloaded by the item context or solution procedures which are required. Semantic-conceptual and operative knowledge approaches allow for a more flexible investigation of subjects' behaviour. Individual perception of problems, thought processes and especially the modes of intuitive and analytic thought are essential ingredients.

Bibliography

Bauersfeld, H.: 1983, 'Subjektive Erfahrungsbereiche als Grundlage einer Interaktionstheorie des Mathematiklernens und -lehrens', in H. Bauersfeld, H. Bussmann, G. Krummheuer, J.H. Lorenz and J. Voigt (eds.), *Lernen und Lehren von Mathematik*, Aulis Deubner, Köln, 1-56.

Bauersfeld, H.: 1985, 'Ergebnisse und Probleme von Mikroanalysen mathematischen Unterrichts', in W. Dörfler and R. Fischer (eds.): *Empirische Untersuchungen zum Lehren und Lernen von Mathematik*, B.G. Teubner, Stuttgart, 7-26.

Bauersfeld, H.: 1988, 'Interaction, Construction and Knowledge - Alternative Perspectives for Mathematics Education', in D.E Grows and T.J. Cooney (eds.), *Effective Mathematics Teaching*, Research Agenda for Mathematics Education, vol.1, National Council of Teachers of Mathematics, Reston, VA, 27-46.

Bauersfeld, H.: 1991, 'The Structuring of the Structures - Development and Function of Mathematizing as a Social Practice', in L.P. Steffe (ed.), *Constructivism and Education*, Lawrence Erlbaum, Hillsdale, NJ.

Borovcnik, M.: 1991, 'Curricular Developments in German Speaking Countries', in D. Vere-Jones (ed.): *Proc. of Third Intern. Conf. on Teaching Statistics*, International Statistical Institute, Voorburg.

Bromme, et al. (eds.): 1981, *Perspektiven für die Ausbildung des Mathematiklehrers*, IDM-Reihe Untersuchungen zum Mathematikunterricht, vol. 2, Aulis Deubner, Köln.

Dawydow, W.: 1977, *Arten der Verallgemeinerung im Unterricht*, Volk und Wissen, Berlin.

Dörfler, W.: 1984, 'Actions as a Means for Acquiring Mathematical Concepts', in *Proc. Eighth Intern. Conf. for the Psychology of Mathematics Education*, Sydney, 172-180.

Fischer R.: 1984, 'Offene Mathematik und Visualisierung', *mathematica didactica* **7**, 139-160.

Fischer, R.: 1988, 'Didactics, Mathematics, and Communication', *For the Learning of Mathematics* **8** (2), 20-30.

Fischer, R.: 1989, 'Social Change and Mathematics', in W. Blum, J.S. Berry, R. Biehler, I.D. Huntley, G. Kaiser-Messmer, and L. Profke (eds.), *Applications and Modelling in Learning and Teaching Mathematics*, Ellis Horwood, Chicester, 12-21.

Fischer, R. and G. Malle: 1985, *Mensch und Mathematik*, Bibliographisches Institut, Mannheim.

Fischbein, E.: 1975, *The Intuitive Sources of Probabilistic Thinking in Children*, Reidel, Dordrecht.

Fischbein, E.: 1977, 'Image and Concept in Learning Mathematics', *Educational Studies in Mathematics* **8**, 153-165.

Fischbein, E.: 1987, *Intuition in Science and Mathematics. An Educational Approach*, Reidel, Dordrecht.

Freudenthal, H.: 1983, *Didactical Phenomenology of Mathematical Structures*, Reidel, Dordrecht.

Hacking, I.: 1990, *The Taming of Chance*, Cambridge University Press, Cambridge.

Konold, C.: in press, 'Understanding Students' Beliefs about Probability', in E. v. Glasersfeld (ed.), *Radical Constructivism in Mathematics Education*, Kluwer, Dordrecht.

Malle, G.: 1983, *Untersuchungen zum Zahlbegriff*, unpublished Habilitationsschrift, University of Klagenfurt.

Moore, D.: 1990, 'Uncertainty', in L. Steen (ed.), *On the Shoulders of Giants: New Approaches to Numeracy*, National Research Council, Washington, D.C.

Piaget, J.: 1975, *Die Enwicklung des Erkennens I: Das mathematische Denken*, Klett, Stuttgart.

Piaget, J. and B. Inhelder: 1951, *The Origin of the Idea of Chance in Children*, English translation 1975, Routledge and Kegan Paul, London.

Wittmann, E.: 1981, 'The Complementary Roles of Intuitive and Reflective Thinking in Mathematics Teaching', *Educational Studies in Mathematics* **12**, 389-397.

Ramesh Kapadia
9 Beechwood Close
Surbiton
Surrey KT6 6PF
U.K.

Manfred Borovcnik
Institut für Mathematik
Universität Klagenfurt
Sterneckstraße 15
A-9020 Klagenfurt
Austria

Chapter 2

A Probabilistic Perspective

M. Borovcnik, H.-J. Bentz and R. Kapadia

There are unusual features in the conceptual development of probability in comparison to other mathematical theories such as geometry or arithmetic. A mathematical approach only began to emerge rather late, about three centuries ago, long after man's first experiences of chance occurrences. A large number of paradoxes accompanied the emergence of concepts indicating the disparity between intuitions and formal approaches within the sometimes difficult conceptual development. A particular problem had been to abandon the endeavour to formalize one specific interpretation and concentrate on studying the structure of probability. Eventually, a sound mathematical foundation was only published in 1933 but this has not clarified the nature of probability. There are still a number of quite distinctive philosophical approaches which arouse controversy to this day. In this part of the book all these aspects are discussed in order to present a mathematical or probabilistic perspective. The scene is set by presenting the philosophical background in conjunction with historical development; the mathematical framework offers a current viewpoint while the paradoxes illuminate the probabilistic ideas.

1. History and Philosophy

Though mathematics is logically structured, its progress has not always been smooth. The history of mathematics has revealed many interesting ideas and paradoxes and has witnessed fierce philosophical controversies. There are some useful sources for the history of probability like David (1962), Maistrov (1974), and Schneider (1989), while philosophical issues are dealt with in Fine (1973) and Barnett (1973).

Tardy conceptualization of probability

Traces of probability can be found in the ancient cultures of Indians, Babylonians, and Egyptians. The earliest known object used for games of chance around 3500 B.C., was

R. Kapadia and M. Borovcnik (eds.), Chance Encounters: Probability in Education, 27–71.

the astragalus, a bone in the heel of a sheep. Betting games with these bones were popular folklore among Roman soldiers. It is possible that primitive dice were made by rubbing the round sides of the astragalus until they were approximately flat. Yet, so long as real bones were used, the regularity of the fall would be obscured. It would undoubtedly be affected by the kind of animal bone used and by its wear.

But the cubes made from well-fired buff pottery which were in use in Babylon 3000 B.C. were nearly perfect dice. Considerable experience would have been gained from casting dice or drawing beans out of urns for divine judgement at religious ceremonies (at Delphi and elsewhere); it is curious that the conceptual breakthrough based on the regularity of the fall of dice did not occur before the birth of Christ. It could be that priests were taught to manipulate the fall of the dice to achieve a desired result, as the interpretation of divine intent. Moreover, speculation on such a subject might have brought a charge of impiety in the attempt to penetrate the mysteries of the deity.

In her book on the origins and history of probabilistic ideas from ancient times to the Newtonian era, David (1962, p.21) wonders why the conceptual progress was so tardy:

> "It was fifteen hundred years after the idealization of the solid figure before we have the first stirrings, and four hundred and fifty years after that before there was the final breakaway from the games of chance, if indeed we have really accomplished it even today."

With the advent of Christianity the concept of the random event of the pagan philosophers was finally rejected: according to St. Augustine everything is minutely controlled by the Will of God. David (1962, p.26) suggests that the step to conceptualize probability did not come at that time because the philosophical development engendered a habit of mind which made impossible the construction of theoretical hypotheses from empirical data.

> "If events appear to occur at random, that is because of the ignorance of man and not in the nature of events."

However, one should not overestimate this tardiness of conceptualization as this is also true for other disciplines. The development of a causal physical approach to science instead of a deistic one was marked by great controversies, even if today one finds such ideas naturally accepted. Think of Galilei's (1564-1642) troubles with the Pope. Euclidean geometry was not the breakthrough as is generally viewed, neither in axiomatic thinking nor in geometry. It only provided rules for construction but no formal concepts; the status of the parallel axiom was not clarified until the work of Gauss and Lobachevski in the nineteenth century; the concept of continuity of lines and planes was developed even later, though it is implicitly assumed by Euclid. In arithmetic, on the other hand, no sound answer was given to the question of axiomatizing numbers until Peano over 100 years ago. However, a comparable milestone was not reached in probability until 1933.

At this point we want to outline another feature of stochastics which may have been an obstacle to developing formal concepts of probability early in history. Faced with a situation of uncertainty, ruled by coins or by divine judgements, the *next* explicit outcome is of major interest. No definite prediction can be given (by humans). Thus one might speculate on patterns of previous outcomes or on divine will. To achieve progress in conceptualization one has to see the next outcome only as a representative of future (or hypothetical) outcomes. Only this transformation of the problem makes it tractable but does not give an answer to the original question. For a coin, the probability of 1/2 tells nothing about which face will actually show up and if 'tail' will occur at the next throw of the coin. However, surprisingly, this probability constitutes some indirect knowledge for this specific toss. This aspect of probability is a major hindrance for children's understanding and is still an issue of philosophical debate (Stegmüller, 1973, pp.98).

The complexity of probability is apparent from a modern perspective but may account for its tardy history. Fine (1973, p.7) classifies probability theories according to five dimensions which are interdependent: the domain of application (which phenomena to be modelled), the form of probability statements (qualitative or quantitative), the relations between probability statements (the calculus), the procedures in measuring (initial) probabilities, the goals of the theory (which intuitive ideas to be formalized). He goes on to characterize eleven different theories of probability using the following terms:
Empirical: physical properties of the world, based solely on experiment and observation.
Logical: formalized thinking, either deductive or inductive.
Objective: interpersonal, independent of thought
Pragmatic: emphasizing the practical and good rather than the true or correct.
Subjective: involving individual opinions, judgements, and experience.

The rule of 'favourable to possible'

Returning to historical development, David attributes Cardano with the first reference to probability. Cardano discusses the cast of one die in *Liber de ludo aleae* (ca. 1560):

> "One half the total number of faces always represent equality; thus the chances are equal that a given point will turn up in three throws, for the total circuit is

> completed in six, or again that one of three given points will turn up in one throw. For example I can as easily throw one, three or five as two, four or six. The wagers therefore are laid in accordance with this equality if the die is honest."

By "the chances are equal that a given point will turn up in three throws ..." Cardano refers to what nowadays is expressed by 'expectation'. Of course the probability of (at least) one given point in a series of three throws is 91/216 which is smaller than 0.5. Cardano's argument is an intermixture of equiprobability and expectation. In 1558 Peverone had analyzed a game of chance by trying to enumerate a probability set, but he did not stick to his own rules; Kendall (1956) calls this one of the near misses of history. On the other hand David (1973, p.58) is certain about Cardano:

> "There is no doubt about it: here the abstraction from empiricism to theoretical concept is made and, as far as I know, for the first time."

Yet one should remember that Cardano simply evaluates a specific probability; it is doubtful that he really made an abstraction from empirical frequencies to a theoretical concept since he makes no attempt to define the concept explicitly.

It was a century later that Pascal and Fermat made great progress in conceptualizing probability in their famous correspondence in 1654 which was not published till 1679. They solved two specific problems, de Méré's problem and the Division of Stakes (problem of points).

De Méré's problem. The favourability of two games is to be compared. In game 1 the player wins if there is at least one 'six' in four throws of a die; in game 2 the player wins if there is at least one 'double six' in 24 throws of two dice. Pascal and Fermat's solution was based on an exhaustive enumeration of the fundamental probability set, following the work of Galilei:

$$P(\text{win in game 1}) = 1 - (5/6)^4 = 671/1296 = 0.508 > 1/2$$

$$P(\text{win in game 2}) = 1 - (35/36)^{24} = 0.491 < 1/2$$

De Méré rejected that solution (Ore, 1953, p.411):

> "If one wants to throw a six with one die, one has an advantage in four throws as the odds are 671 to 625. If one throws two sixes with two dice, there is a disadvantage in having only 24 throws. However, 24 to 36 (the number of cases for two dice) is as four to six (the number of cases on one die)."

It is probabilistic folklore that de Méré won by game 1 and lost all his fortunes by game 2. However, we do not believe that gamblers in the 17th century did observe a difference in these probabilities despite their huge amount of practical experience: a carefully observed sample of 5,000 trials is required to statistically detect such small differences. De Méré's rejection is perhaps based on a kind of theoretical conflict between a direct enumeration of the fundamental probability set and the favourable to possible rule. Now this rule yields 'correct' solutions if applied to favourable events which are single elements of the sample space. But in the games above, the rule is applied to a series of 4 or 24 trials as favourable cases which are not elements of the same sample space. Also, a confusion with expectation occurs in these games. The expected number of 'sixes' in a series of 4 trials is 4·(1/6); for 'double sixes' in a series of 24 it is 24·(1/36). Thus, the expected number of successes is equal in both cases.

Division of stakes. The problem deals with the fair division of stakes if a series of games has to be stopped before completion. At the beginning of a series of games two players *A* and *B* bet equal stakes. The player to win a certain number of single games first wins the whole stake. However, the series has to be interrupted before one of the players has reached the required number of points and the stakes have to be divided. If five games are required to win, and the score is 4:3 in favour of *A*, what is the fair division of stakes? This is an old and famous problem since the 13th century, many non stochastic approaches to it like Pacioli (1494) are known.

Pascal and Fermat's achievement in the problem of points lies in their special approach. They were the first to model the fair division of stakes by a game of chance. They introduced the scenario of what should happen if the game was continued and if the chances of the players were equal for a single round. The stakes should then be divided proportionally to the probability of winning in this continuation of games. Pascal developed his arithmetic triangle as a general method to solve similar problems.

Pascal and Fermat's approach sheds light on the correct application of the favourable to possible rule but they made no progress in defining a concept to clarify the nature of probability. They used probability pragmatically; the equal likelihood of outcomes in games of chance seemed to be intuitively obvious to them. Games of chance served as a link between intuition and developing concepts as well as a tool to structure real phenomena. This view is also supported by Maistrov (1973, p.48):

> "...actual gaming was, as a rule, condemned. These games did serve as convenient and readily understandable schema for handy illustration of various probabilistic propositions."

Expectation and frequentist applications

According to David, the real begetter of the calculus of probability was born in the Hague in 1629. Certainly, calculations of probability were done by Cardano and Galilei while Pascal and Fermat explored a number of interesting problems. But the man to synthesize ideas in a systematic way, with the generalizations emerging from the solutions to problems was Christian Huygens. He derived the rules and first made definitive the idea of mathematical expectation; he put probability on a sound footing. His book *De ratiociniis in aleae ludo* was published in 1657 and was not superseded for over half a century. Indeed parts of his work were incorporated into James Bernoulli's masterpiece *Ars conjectandi* (1713). Huygens (1657, p.66) used expectation and not probability as the central concept:

> "Proposition 3: to have p chances of obtaining a and q of obtaining b, chances being equal, is worth $(pa+qb)/(p+q)$."

Huygens has been vindicated in his choice of this concept by the fact that in practical situations, calculation of arithmetic means is considered appropriate. In fact, Huygens did not speak of expectation, a denomination which stems from the Latin translation of his book by v. Schooten. Instead, he spoke of the true value of the pay-off table. Probability itself plays the role of an undefined elementary concept, which is justified by reference to actual games of chance. Huygens does not develop any combinatorial method; his concept of expectation, recursively applied, yields solutions to many known problems. A branch of applications of probability emerged at that time. Huygens established mortality tables and treated frequencies in just the same way as probabilities. Moreover, he defined theoretical concepts like the mean life time.

The first Englishman to calculate empirical probabilities on any scale was Graunt. This was a natural extension of the Domesday Book, the most remarkable administrative document of its age, which enabled the King to know the landed wealth of his entire realm, how it was peopled and with what sort of men, what their rights were and how much they were worth. Speculations and research into probability theory did not, apparently, concern the English, whose interest seemed to be in concrete facts. Graunt's importance both as a statistician and an empirical probabilist lies in his attempts (1662) to enumerate as a fundamental probability set, the population of London at risk to the several diseases such as are given in the Bills of Mortality recorded by London clerks. This gave further impetus to the collection of vital statistics and life tables. As Graunt epitomizes, the empirical approach of the English to probability was not through the gaming table, but through the raw material of experience.

The conceptual progress, however, underpinning the justification for the link between relative frequencies and probabilities was not mastered before James Bernoulli (1713). He worked for over 20 years on the law of large numbers which he called 'theorema aureum'. This theorem proves that the relative frequencies, in some sense, converge to the underlying probability. It justifies the use of probabilities in contexts outside games. Bernoulli also marks a shift from Huygen's expectation to the probability concept as his book gives a systematic treatment of combinatorics. His philosophical ideas, however, are still of metaphysical determinism. Everything is ruled by deterministic laws - weather, dice or eclipses of planets likewise. Chance only occurs because of our limited knowledge of these laws, so the applications of current probability would be restricted to games. Most phenomena would be so complex that it would be pointless to study the possible cases. However, after having proved his theorem, he can state that this is

> "... another way that will lead us to what we are looking for and enable us to at least ascertain a posteriori what we cannot determine a priori, that is, to ascertain it from the results observed in numerous similar instances" (from Maistrov, 1974 p.69).

Inference and the normal law

It is remarkable that there are traces of statistical inference before any attempt to define the probability concept. Arbuthnot (1710) analyzed the birth statistics of London for 80 successive years and found that there were more boys than girls born in each of the years. If God does not exist, he argued, then there is no special cause whence the probability of a child to be a boy or a girl is equal. He concluded that the probability for more boys than girls in a single year is 1/2 so that the chance of more boys than girls in 80 consecutive years has a probability of $(1/2)^{80}$, a quantity with 24 leading zeros after the decimal point. As this probability is extremely small, the empirical data were taken to confirm 'God's will in action'. A similar argument was used by Buffon (1740) to prove that the planets originated from a common cause which was thought to be a collision of the sun with a comet. The type of argument used by Arbuthnot and Buffon is now called likelihood argument. According to that a hypothesis H is evaluated by means of an actually observed event E via the conditional probability $P(E|H)$. If this quantity is small, then the hypothesis H is rejected. Today the argument is not used to evaluate a single hypothesis but only to compare the plausibility of competing hypotheses (Stegmüller, 1973, p.145).

D. Bernoulli (1777) used a maximum likelihood argument to prove that the best choice from a series of observations can differ from the arithmetic mean by using a particular continuous distribution of errors. However, he did not calculate a probability for the error to exceed a specific quantity but looked for a best choice to represent the data. His argument can be summarized in the following form. A certain event E has occurred which has different probabilities under different, mutually exclusive conditions C_i. Then one should assume that condition being operative which supplies the observed event with the highest conditional probability $P(E|C_i)$. This is clearly a precursor to R. A. Fisher's maximum likelihood principle.

Bayes (1763) derived what we now call a distribution for the unknown parameter p in a Bernoulli series based on relative frequencies k/n from which he could calculate interval estimates for p. This is not to be confused with classical confidence intervals as the unknown parameter p is now a random variable which has its own probability distribution before and after the sample is drawn. In his apparatus for simulating the Bernoulli series, the initial uniform distribution of p on the interval (0,1) was apparent. Even his generalizing argument for an equidistribution for this binomial parameter has some objective rationale (Edwards, 1978) and hence differs from a modern subjectivist view.

There are some indications of the concept of probability distribution emerging in the 18th century. De Moivre (1738) was the first to find the function which is now called the normal density. He looked for an approximation of sums of binomial probabilities in studying the deviations in the Bernoulli theorem. In modern terms, he looked for a distribution of deviations of the relative frequency H/n from the underlying probability p for a fixed sample size n. He solved this for $p=1/2$ by deriving the limiting distribution for $n \to \infty$. For de Moivre, the normal density served only as a tool for numerical approximation and had no proper probabilistic meaning. It was not before Simpson (1755) that continuous distributions were used at all. Simpson modelled the fluctuation of observations by a triangular distribution, and thus gave an argument for the use of the mean instead of a whole series of measurements.

It was Laplace's most important statistical achievement to derive the central limit theorem (1810) which, in essence, states that the binomial distribution approaches the normal distribution as the number of trials increases to infinity. Laplace also developed an intuitive idea when to apply this normal distribution; whenever a huge number of quantities add up to a specific (observable) variable, this latter variable will be governed by a normal probability law. He believed that it could play an analogous role to the law of universal gravitation by which most celestial phenomena are explained. Any general variable could be explained by the unique normal law by splitting it into a sum of additive quantities, the distributions of which may even remain unknown.

Gauss (1809) used the normal distribution not only as a tool for approximation but as a distribution in its own right. His approach was embedded in error theory; in deriving the mean as the most likely of values to replace several repeated measurements of an unknown quantity, he recognized that one had to know the distribution for the measurement errors first. As this error distribution was not known, he reversed the argument. Which hypothetical error distribution would yield the mean as the most likely value from a series of measurements? Gauss took the mean as given, and found the normal distribution as one base to derive the mean as most likely value. He also gave intuitively acceptable mathematical restrictions from which he could derive the distribution of errors to be normal.

Gauss explored the relationship between four concepts; the mean as the best value to take from a series of measurements; the normal distribution for describing variation of errors; the maximum likelihood method to take the best value from a series of measurements; and the method of least squares to derive the best value replacing a series of measurements. Assuming some of these concepts were true, the others could be derived. So, for example, Gauss could prove that the methods of maximum likelihood and least squares are equivalent for the normal distribution. Overall, the interrelations between these concepts justified all of them.

Laplace's intuitive argument on the universality of the normal distribution was soon picked up by others; Bessel (1818) tried to give a practical proof of these results by experiments; Quetelet (1834) applied this error distribution to biometric measurements. He developed the idea of the average man in analogy to the error theory. This l'homme moyen, in his terminology, is what is ideally to be measured with reference to a specific variable, say the circumference of the breast. The single man to be measured will differ from this fictitious mean by errors of nature. These errors are assumed to be of the same type as measurement errors, especially as they are the additive resultant of 'elementary errors'. Biometric measurement on a large scale followed and, in many cases, the normal distribution was found to govern many phenomena. Galton (1889) constructed the Quincunx to simulate a practical demonstration of binomial trials and thereby the central limit theorem. There was a romantic enthusiasm which may best be expressed by Galton's own words (1889, p.66):

> "I know of scarcely anything so apt to impress the imagination as the wonderful form of cosmic order expressed by the 'Law of Frequency of Error'. The law

> would have been personified by the Greeks and deified, if they had known of it. It reigns with serenity and in complete self-effacement amidst the wildest confusion. The huger the mob, and the greater the apparent anarchy the more perfect is its sway. It is the supreme law of Unreason."

With this enthusiasm in mind, the title of 'the normal law' is easier to understand. The prominent role of this law did not change even after other distributions like the Maxwell distribution became of interest in physics and K. Pearson (1902) investigated other types of continuous distributions systematically. The Russian school (Chebyshev, Markov) pursued various generalizations of the central limit theorem bringing in the idea of measure theory. A decrease in importance is to be noted only these days with the advent of robust statistics, nonparametric statistics, and exploratory data analysis.

Foundations and obstacles

We now return to the foundations of probability where Laplace's work marked a culmination and final stage in the early conceptual development. Philosophically, however, his views were still based on a mechanical determinism.

> "Given ... an intelligence which would comprehend all ... for it, nothing would be uncertain and the future, as the past, would be present to its eyes ... Probability is relative in part to (our) ignorance, in part to our knowledge." (1814,1951, p.4,6).

Laplace gave the first explicit definition of probability, the so called classical probability. The probability P(A) of an event A equals the ratio of the number of all outcomes which are favourable to event A to the number of all possible outcomes of the trial. This definition implicitly assumes that the individual outcomes are equally likely. Laplace formulated the 'principle of insufficient reason' to check this assumption. According to that principle we should assume outcomes to be equally likely if we have no reason to believe that one or another outcome is more likely to arise.

This first formal definition could not clarify the nature of probability, as it refers to a philosophically obscure principle for its application and as its domain was far from actual applications then. Ensuing attempts to amend this principle (by indifference or invariance considerations) have not been successful: according to Barnett (1973, p.71) they "cannot stand up to critical scrutiny". There were also some problems and ideas which hindered conceptual progress.

The St. Petersburg paradox is a problem in which a player has to pay an infinite amount of money as a fair stake, which is unacceptable. To solve the paradox, D. Bernoulli proposed to average the utilities of payments and not the specific payments. He assumed the utility to be a logarithmic function of payments and derived a finite and intuitively acceptable *moral expectation* as fair stakes. The concept of moral expectation gained much support from Buffon, Condorcet, Laplace, or Poisson but the time was not ripe to consider the special choice of the utility function merely as one of several possible models. In the course of his likelihood argument for the solar system, Buffon needed a rule for deciding which likelihood is small enough to establish the 'proof' of a common cause. For practical purposes, an event with a probability as small as 0.0001 should be considered impossible and its complement as certain. Such ambiguous rules about *moral certainty* can also be found with other writers like Cournot.

Independence was a steady source of difficulty as it was not dealt with as a distinct concept. D'Alembert (1754) opposed the equiprobability of the four outcomes in tossing two coins; he argued fiercely in favour of the probabilities 1/3 for each of the outcomes 0,1, and 2 'heads'. His arguments were not countered by reference to the independence concept. Moreover, in calculating the probability of the judgement of tribunals to be correct, the error probabilities of single judges were multiplied as if they were independent. If the tribunal consisted of 7 judges and each of them were right with probability 0.97, then the probability of the majority being right was calculated to be 0.99998, which is 'morally certain'.

Careless use of *Bayes' formula* leads to the conclusion that any conjecture has the probability 0.5 of being true if one knows nothing about it. This stems from the following line of argument. One repeatedly observes a phenomenon like a Bernoulli series with unknown probability p of a specific event occurring. At the beginning of the observations one knows nothing about the phenomenon; then in n observations, the event occurs k times. If the initial ignorance of the value of p is modelled by a uniform distribution (p is treated like a random variable), then according to Bayes' formula, the probability of the event occurring in the next observation is calculated to be $(k+1)/(n+2)$. It may be strange to say that the probability of the sun rising tomorrow ever increases the longer one has observed it to rise. However, for the special case of $n=0$ and $k=0$ when there is no data or knowledge about any outcomes of trials, the probability for the event occurring equals 1/2. This clearly is paradoxical as there cannot be a general rule transforming complete ignorance into a probability statement of 1/2 which establishes some kind of information. However, the root of this trouble is basic misunderstanding between objectivist and subjectivist concepts.

Laplace's definition necessitated a rule to guide its application. Bayes had justified the use of a uniform distribution on the interval (0,1) for a binomial parameter p in the face of complete ignorance of the outcomes of the Bernoulli experiment. His argument was correct but not completely understood; it was transferred by Laplace to formulate the 'principle of insufficient reason' as a basic guide to apply his probability definition. In the extreme case of complete *ignorance*, Laplace's principle could still be used to derive an equidistribution on all possible states, which clearly is a kind of *information*. If taken for granted, this principle thus would yield another ambiguous rule to transfer ignorance into knowledge.

The attempt to *represent* the *equally likely* cases in a *truly objective* way causes difficulties. One way to justify Laplace's principle is to look for *physical* symmetries of the random phenomenon in question, e.g. a physical symmetry of the die should directly lead to equal likelihood of its faces. However, there are many plausible physical symmetries, so a truly objective theory requires a procedure for choosing a particular symmetry and justifying that choice. Indeed, Fine (1973, p.167) illustrates the difficulties by the example of scores with two dice, where there are at least three models.

Maxwell-Boltzmann model: each of the 36 pairs (1,1), (1,2), .., (6,5), (6,6), is equally likely, so that pairs like (2,3) and (3,2) are treated as different outcomes.
Bose-Einstein model: each of 24 pairs (1,1), (1,2), .., (1,6), (2,2), (2,3), .., (5,5), (5,6), (6,6), is equally likely, so that pairs like (2,3) and (3,2) are treated as identical outcomes.
Fermi-Dirac model: each of 18 pairs (1,2), (1,3), .., (1,6), (2,3), (2,4), .., (5,6) is equally likely, so that the two components are forbidden to have the same value.

For ordinary dice, the Maxwell-Boltzmann statistics is the natural model. The two dice are discernible; blue and red dice, or first and second trial; the independence assumption is highly plausible. This natural model, however, is not true for many applications in physics. According to Feller (1957, p.39), numerous attempts have been made

> "to prove that physical particles behave in accordance with Maxwell-Boltzmann statistics but modern theory has shown beyond doubt that this statistics does not apply to any known particles."

The Bose-Einstein statistics, is found to be appropriate for photons, nuclei and atoms containing an even number of elementary particles which are essentially indistinguishable. The Fermi-Dirac statistics is appropriate for electrons, protons and neutrons where the particles are essentially indistinguishable and each state can only contain a single particle. It is startling that the world is found to work in this way, experimentally. One is reminded of the similar situation where non-Euclidean geometry is found to be a better model of the world than Euclidean geometry which reigned supreme and unchallenged for over two millennia.

Thus the natural model for dice and its underlying symmetry is often inappropriate for physical particles which considerably reduces its scope. The example also shows that there are *different* possible symmetries in *one* physical situation; it is interesting to note that pedagogical problems also arise over this issue. Therefore, probability cannot be an inherent feature of real objects but only a result of our endeavour to model reality. The intention to derive unique probabilities has led to indissoluble paradoxes. Fine (1973, p.167) gives many counterexamples and summarizes his critique against Laplace's definition:

> "We cannot extract information (a probability distribution) from ignorance; the principle is ambiguous and applications often result in inconsistencies; and the classical approach to probability is neither an objective nor an empirical theory."

Axiomatization of probability

Probability theory gained a substantial conceptual role in physics, particularly in thermodynamics, during the last decades of the 19th century. Some new physical laws could only be described in probabilistic terms (like second main law of thermodynamics). Statistical applications at the same time culminated in biometric developments, especially in the 'invention' of regression and correlation. Probability thereby was treated mostly as an empirical concept (as relative frequency), and in some respects also as a theoretical concept (as entropy in physics). This development is countervailed by the fact that there was still no adequate foundation, the only existing attempt was that by Laplace which had its disadvantages. At the 1900 mathematical congress in Paris, David Hilbert formulated a programme for mathematical research among which a satisfactory axiomatization of probability and statistical mechanics was one of the major tasks.

R. v. Mises (1919) was one of the pioneers in axiomatization. The interpretation of probability as relative frequency and the convergence of relative frequencies in Bernoulli's theorem were the basis for his research. Fine (1973, p.90) summarizes the frequentist position as follows:

(1) In many important cases relative frequencies appear to converge or stabilize when one sufficiently repeats a random experiment.

(2) Apparent convergence of relative frequency is an empirical fact and a striking instance of order in chaos.

(3) The apparent convergence suggests the hypothesis of convergence and thereby that we may extrapolate from observed relative frequency to the relative frequency of outcomes in as yet unperformed trials of an experiment.

(4) Probability can be interpreted through the limit of relative frequency and assessed from relative frequency data.

However, this approach was not successful; it was too complicated and philosophical troubles were overwhelming. Certain critical points were clarified only recently (see C.P. Schnorr, 1971). For instance, the Bernoulli theorem of frequencies converging to the underlying probability does not involve the usual convergence but is, in itself, based on probabilities. Either these probabilities are of the same type and hence must not be used for the defining relation, or they are of a different type, which then needs to be clarified and defined. Furthermore, von Mises' definition was based on a randomness property of sequences which is a philosophically obscure concept and difficult to check in any case of application. The only way to clarify this randomness property is by means of probability itself which again marks a circularity in von Mises' approach.

It was Kolmogorov (1933) who finally formulated a system of axioms for probability and derived the usual theorems which were immediately acknowledged. His approach signifies the culmination of measure theoretic arguments which had gained importance in proving various generalizations of the central limit theorem. However, Kolmogorov's approach did not really clarify what probability is; it only elaborated the *structural* properties of probability and left the interpretation of the thereby defined concept of probability as an open question. It is to be noted that, however, this approach was primarily thought of as a justification of the frequentist interpretation of probability. Despite this successful axiomatization, the real controversy in foundation between subjectivists and objectivists revived later in the 1930s with Jeffreys (1939) and de Finetti (1937).

Modern views on probability

One begins to appreciate the multi-faceted character of probability even after the universally accepted foundation by Kolmogorov, when one looks at the various classifications of probabilities by Fine (1973). We will discuss only the four main approaches to the nature of probability which are relevant for school mathematics. They are summarized below as classical, frequentist, subjectivist and structural.

Classical view. According to Laplace the probability of a combined event is obtained by the fraction of outcomes favourable to this event in the sample space; this makes use of an implicit assumption of equal likelihood of all single outcomes of the sample space. It is an a priori approach to probability in that it allows calculation of probabilities *before* any trial is made. Geometric probability is closely related to it; it reduces the probability concept to counting or area.

In the case of applications, one is confronted with the problem of deciding which are the single outcomes that are equally likely. Symmetry in the physical experiment with respect to Laplace's 'principle of insufficient reason' is a shaky guideline to help in this respect. One major philosophical problem is that the same physical experiment can reveal several different symmetries.

Frequentist view. The probability of an event is obtained from the observed relative frequency of that event in repeated trials. Probabilities are not obtained exactly by this procedure but are estimated. It is an a posteriori, experimental approach based on information *after* actual trials have been done. The measure of uncertainty is assigned to an individual event by embedding it into a collective - an infinite class of 'similar' events which are assumed to have certain 'randomness' properties. Then the probability is the limit towards which the relative frequency tends.

In applying this definition, one is faced with the problem that an individual event can be embedded in different collectives, with no guarantee of the same resulting limiting frequencies: one requires a procedure to justify the choice of a particular embedding sequence. Furthermore there are obvious difficulties in defining what is meant by 'similar' or by 'randomness'; indeed an element of circularity is involved. Even the notion of settling down presents difficulties in terms of the number of trials needed in long term frequency.

Subjectivist view. Probabilities are evaluations of situations which are inherent in the subject's mind - not features of the real world around us which is implicitly assumed in the first two approaches above. The basic assumption here is that subjects have their own probabilities which are derived from an implicit preference pattern between decisions. Ideas are borrowed from gambling where given any event, its probability can be determined by the odds a person is prepared to accept in betting on its occurrence. Obviously people might differ in the odds they would accept, but this does not matter provided basic rules of coherence and consistency are met. For example, it would be foolish to place two bets of 3 to 2 on two horses in a two horse race because one is bound to lose money as the win of £2 on one does not compensate for the loss of £3 on the other. Coherence formalizes this basic idea from which one can deduce the basic laws of probability (Barnett, 1973).

For a subjectivist there are two categories of information, namely prior information which is independent of any empirical data in a subject's mind, and empirical data which amount to frequencies in repeated experiments. Both types of information are combined in Bayes' formula to give a new probability of the event in question. This updating of probabilities is called 'learning from experience'. Bayes' formula combined with simple criteria of rationality allow a *direct* link from experience and prior information to decisions. Thus, the usual problems of statistical methods based on objectivist probability are circumvented.

A problem inherent to the subjectivist approach is its intended ubiquity; any uncertainty has to be specified by probabilities. There might be occasions when it is better not to force meagre information into the detailed form of a distribution; it might be wiser to try a completely different method to solve the problem. The most striking argument against the subjectivist position, however, is that it gives no guidance on how to measure prior probabilities (Barnett, 1973, p.227). Though there may be flaws in the classical and frequentist approaches, they do provide pragmatic procedures to calculate probabilities. Of course, a subjectivist would exploit all frequentist and symmetry information to end up with a suitable prior distribution. Much effort from subjectivists is devoted to develop strategies to improve the measurement of prior probabilities.

Structural approach. Formal probability is a concept which is implicitly defined by a system of axioms and the body of definitions and theorems which may be derived from these axioms. Probabilities are derived from other probabilities according to mathematical theorems, with no justification for their numerical values in any case of application. This structural approach does not clarify the nature of probability itself though the theorems derived are an indicator of possible interpretations. The structural approach, however, may serve as a theoretical framework for the two main conceptions of probability.

The *objectivist* position encompasses the classical and the frequentist view; according to it, probability is a kind of disposition of certain physical systems which is indirectly related to empirical frequencies. This relation is confirmed by theorems like the law of large numbers. The *subjectivist* view treats probability as a degree of confidence in uncertain propositions (events). Axioms on rational betting behaviour like coherence and consistency provide rules for probabilities; the Kolmogorov axioms prove mathematical theorems which must be obeyed if one wants to deal rationally with probabilities.

It is remarkable that the structural (probability) theories built on the various systems of axioms are identical between objectivists and subjectivists. At least this is true if one neglects the more radical subjectivist variants like de Finetti, who denies even the σ-additivity of probability measures. That means that the competing positions use precisely the same axiomatic language with respect to syntax. Of course, the semantics is different; Kolmogorov's axioms are not the basic medium for the interpretation between theory and reality, but are logically derived rules. Despite the formal equivalence of the approaches, Kolmogorov's axioms are usually thought to be a justification of the objectivist, especially the frequentist view.

A final example. A standard example of tossing a particular blue die might help to illustrate the various philosophical positions. The classical view would assign a probability of 1/6 of getting a six as there are six faces which one can assume are all equally likely; the frequentist view would assign a probability by doing repeated experiments or, as experiments with other dice have yielded 1/6, the same value would be assigned for the die in question. In both these viewpoints, the probability is an inherent feature of the die, and tossing procedure. The subjectivist would assert that the probability was a mental construct which he/she may alter if new information became available about the die. For example, a value different to 1/6 might be assigned if the die is black or heavier or smaller or was slightly chipped. Clearly there needs to be some basis for the decision made. However, since the probability is not viewed as inherent in the object this would not involve any logical problems because of the conflicting values. In fact the subjectivist does not reject considerations of symmetry or frequency - both are important in evaluating probabilities. There is an insistence, however, that ideas of symmetry or frequency are made explicit and used with care.

The structural view does not help in determining a value for the probability. In any case of application one has to choose a specific subjective or objective interpretation to determine the model and inherent probability. Philosophically, for the structuralist, there is no way of deciding which model is better. This controversy on the nature of probability will continue. Fine (1973) states that

> "subjective probability holds the best position with respect to the value of probability conclusions, ... Unfortunately, the measurement problem in subjective probability is sizeable and conceivably insurmountable ... The conflict between human capabilities and the norms of subjective probability often makes the measurement of subjective probability very difficult." (p.240)

The historical and philosophical background provides a perspective for teachers in deciding how to present ideas to pupils and students.

2. The Mathematical Background[+]

Historically the development of probability has been rather chequered from vague, intuitive notions to precise mathematical formulations. Philosophical controversies are unresolved; there are deep divergences in opinion on the nature of probability between the various schools of probability. Nevertheless, a specific body of mathematics is relatively uncontentious and can be understood as the common language accepted by most positions. We now present this mathematical background.

Model-building

Science is the process of observing natural phenomena, building abstract models which, according to their own inherent rules, may be expected to behave in a similar manner to what we have observed. We use the models to predict further properties of these phenomena and perform experiments to test the validity of our predictions. If necessary, the models are refined or modified, and so forth. This description applies as well to probability modelling as to any other scientific methodology. The distinguishing feature of a probability model is that it incorporates uncertainty, or random error in a formalized way. This randomness, however, introduces a complication. The test of the validity of models is no longer direct, as it is with deterministic models. Sometimes probabilistic models allow us to predict specific outcomes only with an additional measure of confidence (probability), sometimes behaviour is guided without a prediction of specific outcomes.

Now, if a prediction is not fulfilled, then there are two possibilities: first, the model really does not fit; second, the model does fit but a case outside the confidence limit has occurred. This randomness is naturally inherent with probability and it is the risk of applying probability models. If a prediction is correct in a single case, it may nevertheless stem from an inadequate model. We will illustrate this by an example. Imagine a person who has some model to predict outcomes of a state lottery and predicts correctly this

---

[+] Thanks are due to Dr. David Grey for the formulation of some of the ideas in this section.

week. This success is not a confirmation of this model but more likely to be luck. Because of this success it is hard to find convincing theoretical arguments of the inadequacy of this model. On the other hand, the equidistributional model does not allow for a specific prediction in this case as each outcome is as likely as any other. How can one check the validity of this model in the occurrence of, say, a run of certain numbers in the last month?

Thus, probability models are in some way different from deterministic models with respect to their evaluation in specific real situations, since they are intended to integrate randomness. Despite this reservation, we lose nothing of the intellectual rigour of other disciplines but open up the possibility of applications in areas such as medicine and social sciences where deterministic views are often not appropriate. Our intention is to make accessible simpler theory and mathematical concepts by viewing probability as a model-building device and by developing historical and intuitive ideas.

Assigning probabilities

Probability statements refer to 'experiments', a notion which is interpreted liberally here. The construction of a probability model starts with the description of the set of all possible outcomes of such an experiment. This set is called the *sample space* S. It need only be described in as much detail as is required for the purpose of the model. For instance, if we are interested in the outcome of throwing a fair die, then we shall not normally be concerned with the location or orientation of the die when it lands, but merely with which face it lands uppermost. Hence a suitable sample space is $S = \{1,2,3,4,5,6\}$. Even for such a simple experiment as spinning a pointer and observing where it comes to rest, the sample space is uncountably infinite, namely $S = [0,2\pi]$.

Historically, the concept of sample space was linked to the idea of equiprobability. This seemed natural for games of fortune. However, even within this context this link has been an obstacle as is shown by the difficulties in finding the set of all possible outcomes for two dice. The sample space had to include possibilities which were all equiprobable. Thus, it was not just a simple combinatorial enumeration but the problem of whether to respect or neglect order. A solution arose by elaborating the concept of independence which can be used to combine probabilities from each single die via the multiplication rule.

A particularly useful device for constructing sample spaces is the *Cartesian product* of two or more sets. If S_1 and S_2 denote two sample spaces for two different experiments performed in succession, say, then a suitable sample space for the *combined experiment* is the Cartesian product $S_1 x S_2$. This is the set of all ordered pairs (e_1,e_2) where $e_1 \varepsilon S_1$ and $e_2 \varepsilon S_2$. There is no conceptual problem in extending this construction to more experiments, even to an infinite sequence of experiments.

To separate the concepts of sample space and equidistribution and to define the sample space for combined experiments simply by Cartesian products may now seem natural but was one of the milestones of conceptual development. The critical step was the question of how to expand the probabilities of several experiments to the combined one, which is achieved either by the independence assumption or by special, assumed dependencies. A sample space is not always a description of really possible cases as illustrated by the 'Division of Stakes' in the section on paradoxes.

An *event* is any verbal description of results of the experiment analyzed. It may or may not occur as a result of performing this experiment. It is customary to identify an event mathematically with a subset of the sample space. If the sample space is finite or countably infinite, any subset can be used. If the sample space is uncountably infinite, for technical reasons, it is necessary to restrict the definition of probability to a certain class of subsets. However, any verbal description of practical value will be matched by a suitable subset within this restricted class so that mathematical complications are not a drawback in practice.

On the intuitive side, it is hard to accept that there are subsets of uncountably infinite sample spaces which are not allowed to be events. The mathematical reason is that it is impossible to find a probability function which would assign a value to any subset of the real numbers and yet fulfill all the mathematical requirements of probability. This is more a peculiarity of mathematics and real numbers than a feature of randomness of the phenomenon to be modelled. Historically it was an achievement to find the analogy between probabilities and measures, as the Borel subsets of the reals could serve as a suitable range of definition for axiomatic probability. Thus the structural approach can embrace all the special cases of probability, especially in the form of density functions like the 'normal law'.

Having constructed a sample space, probabilities are assigned to events. From the structural view it does not matter whether these probabilities as numerical values are realistic or not; the model can always be validated later. What matters more is that they are internally consistent. By this, it is universally agreed that they should satisfy *Kolmogorov's axioms*:

A_1 $P(E) \geq 0$ for any event E

A_2 $P(S) = 1$ for the whole sample space S

A_3 If $E_1, E_2,..$ is a sequence of mutually exclusive events,then

$$P(\bigcup_i E_i) = \sum_i P(E_i)$$

The first two conditions mean that probabilities are non-negative and that certainty is characterized by a value of 1. The substantial condition of *additivity* embodied in A_3 means that, mathematically, probability is a measure. It may be regarded in some respect as analogous to 'area', 'mass', or 'weight', measures which also share the additivity property.

Barnett (1973, p.69) describes these axioms as a "mathematical milestone" in laying firm foundations, yet a "philosophical irrelevance" in terms of explaining what probability really is. Kolmogorov (1956, p.1) believes that

> "The theory of probability, as a mathematical discipline, can and should be developed from axioms, in exactly the same way as Geometry and Algebra ... all further exposition must be based exclusively on these axioms, independent of the usual concrete meaning of these elements and their relations."

This is overtly true for a mathematical clarification of the nature of probability and serves as a justification of the discipline. As Barnett stated, however, this approach really is a philosophical irrelevance. For the cognitive reconstruction in the process of learning, we hold the view that intuitions on probability have to be considered seriously if teaching is to be successful.

The structural approach does not clarify what probability really is, although, indirectly, the whole body of theorems indicates how concepts are to be understood and applied. As there is a common core of assumptions upon which an uncontroversial mathematical theory of probability has developed, providing a common syntax between subjectivist and objectivist positions in particular, it is now viewed as a strength rather than a weakness of the structural approach that it is open to several different interpretations. The subjectivist's axioms would be a set of axioms on rational (betting) behaviour in uncertain situations. Within the objectivist position there might also be a foundation on random variables as basic notions and a set of axioms on expectation of random variables; this approach (see Bentz, 1983) would, of course, alter semantics and intuitions, even if it, too, may be specialized to be equivalent to the Kolmogorov axioms.

All the usual interpretations of probability are subsumed by the axiomatic approach: Laplace's definition is but a special case; the frequentist view is covered by theorems like

Bernoulli's law; even subjective views are partially subsumed insofar as their axioms lead to the same syntax for probability. All discrete distributions are special cases of the definition as are probabilities defined by density functions. From a structural perspective, probability is defined to be a function which has specific characteristics like additivity. For applications one has to find the specific values of this function or make additional assumptions about the situation to derive such values.

The special choice of a set of axioms will thus have deep consequences on semantics. The axioms are the foundation of the theory and are simultaneously at the interface between theory and reality. Thus to refer to Kolmogorov's axioms as the basic set will have consequences on how to apply mathematical theory. Furthermore the basic axioms could be considered as models of intuitive ideas of probability to be sharpened by the theory. This might be thought of as the historical genesis of ideas in general and also in the sense of how ideas settle down in an individual's learning.

Conditional probability

What distinguishes the theory based on this set of axioms from measure theory is the concept of independence, which is part of a fundamental definition, but not, interestingly enough, part of the axioms. This independence relation is the key assumption of fundamental theorems like the Bernoulli's law of large numbers. Such theorems established a link from the structural approach to the frequentist interpretation and thus contributed to the immediate acceptance of Kolmogorov's axioms within the scientific community. The probability of an event might change if new information becomes available; this is modelled by the notion of conditional probability. If E and F are two events in a sample space S, and E has a nonzero probability, then the *conditional probability* of F given E is defined by

$$P(F \mid E) = \frac{P(E \cap F)}{P(E)}$$

If E is assumed to have occurred, the probability of F is to be evaluated in a new space, namely $S^* = E$. The analogy to area is helpful. The proportion of F which lies in the set E is simply the area of $E \cap F$ divided by the area of E. Mathematically, $P(F|E)$ for fixed E and varying events F is again a measure of probability, satisfying axioms A_1 - A_3, sometimes denoted $P_E(F)$. The definition of conditional probability underlines the fact that probability is a function of available information where changes in the sample space may change the probability.

If $P(F|E) = P(F)$ then F is said to be *independent* of E. The knowledge that E occurs does not alter the probability of F. This can be shown to be equivalent to the following symmetrical relationship between events

$$P(E \cap F) = P(E)\,P(F).$$

If probability defined by Kolmogorov's axioms is to be interpreted in terms of relative frequencies then it must also allow for modelling repeated experiments. With the concept of independence of two similar trials E_1 and E_2 (e.g. of tossing a first and a second coin), it is possible to derive a probability measure on the combinatorial product space for the two trials, using only the distributions on the single sample spaces. For the example of two coins, we obtain (the subscript denotes the number of trial)

$$P(\{(H,H)\}) = P(H_1 \cap H_2) = P(H_1)\,P(H_2) \quad P(\{(H,T)\}) = P(H_1 \cap T_2) = P(H_1)\,P(T_2)$$
$$P(\{(T,H)\}) = P(T_1 \cap H_2) = P(T_1)\,P(H_2) \quad P(\{(T,T)\}) = P(T_1 \cap T_2) = P(T_1)\,P(T_2)$$

This is why independence becomes the key concept within the Kolmogorov approach; it is precisely that concept which allows for modelling repeated experiments which in turn leads to Bernoulli's theorem and the central limit theorem. The generalization of independence to more than two events involves some mathematical complication in representation but no conceptual difficulties. The representation above was loosely written as H_1 and H_2 are subsets of S_1 and S_2 respectively and the sample space of the combined experiment is the Cartesian product $S = S_1 x S_2$. To find an intersection of H_1 and H_2 these events from single experiments have to be embedded into the product space first.

$H_1^* = H_1 x S_2$, 'head on first, no restriction on second trial',

$H_2^* = S_1 x H_2$, 'head on second, no restriction on first trial'.

Then the intersection is defined and yields $H_1^* \cap H_2^* = H_1 x H_2$.

More complications arise in the case of infinite sample spaces for single trials as the sets of the form $E_1 x E_2$ which are known as *cylinder sets*, are only a small part of all subsets of $S_1 x S_2$. In practice, events in the combined experiment may not be of this special form, e.g., in spinning a pointer twice, consider the event 'position of the trials differ by more than $\pi/4$'. Again, this complication is not truly a conceptual difficulty of probability as a phenomenon to be modelled but part of specific aspects of mathematics. For applications it is fortunate that assigning probabilities to cylinder sets is sufficient to determine uniquely an extension of this assignment to probabilities of all events.

Random variables and distributions

In practice one is often interested in some *numerical* value which is associated with and determined by the observed outcome of the experiment. This is known as a random variable. It is random in the sense that the outcome is determined by chance; however, its value is known once the experiment is performed and the outcome is known. The information needed to model a stochastic phenomenon is the set of all possible outcomes and their associated probabilities.

Example: a) In throwing a die with $S=\{1,2,...,6\}$, the denotation $X=i$ is simply an abbreviation for the event 'the die shows i'. The following function is used to describe the probabilities $P(X=i)$.

$$i \rightarrow \frac{1}{6} \quad \text{for } i = 1, 2, .., 6$$

b) In modelling the lifetime X of electronic bulbs, $X \leq x$ is simply an abbreviation for the event 'lifetime X does not exceed x'. The following functions are used to describe probabilities $P(X \leq x)$; the model is called the exponential distribution.

$$f(t) = \tau e^{-\tau t}, \; t \in \mathbf{R}^{+} \text{ and } \tau > 0$$

$$P(X \leq x) = \int_0^x \tau e^{-\tau t} dt = 1 - e^{-\tau x}$$

These two descriptions of random phenomena involve simple *point* functions to describe probabilities. This is different from axiomatic probability which involves *set* functions P which map events onto the real line:

a) $E \rightarrow P(E)$ = proportion of E in S for all $E \subseteq S=\{1,2,..,6\}$.

b) $E \rightarrow P(E)$ for all subsets E of positive real numbers which are allowed to be events (which are measurable). If E is an interval (a,b), then $P(E)$ is calculated by means of integrals

$$P(E) = \int_a^b \tau e^{-\tau x} dx$$

In case b) one cannot even imagine all the mathematically possible events; many of them are not relevant from a practical viewpoint. The calculations become clumsy and a new type of integral is required. Axiomatic probability is especially designed to derive a body of theorems which *indirectly* contribute to a clarification of the nature of probability. In practice, however, random variables and their distributions are to be preferred.

A *random variable* is defined as a function X from the sample space S with an associated probability P into the real numbers $\mathbf{R}$:

$$X\colon S \to \mathbf{R}$$

Not every function may serve as a random variable. One requirement is that for every interval I, the set $\{s\colon X(s) \in I\}$ is an event in the sample space S and thus has well defined probability. This guarantees that the random variable transports the probability P on the sample space S onto the real line by defining

$$\mathbf{P}_X(I) = \mathbf{P}(\{s\colon X(s) \in I\})$$

The range of values of X is known as the *induced sample space* S_X; the induced probability distribution P_X is called the *distribution* of X. Treating random variables as mathematical functions allows for deriving the distribution of functions like X^2 from that of X. An important application of that is the derivation of the distribution of the arithmetic mean $\overline{X} = (X_1+X_2+...+X_n)/n$ which models the mean of repeated trials.

To reconcile this theoretical approach with practical needs, the common procedure of describing random phenomena by point functions can be interpreted as ignoring the underlying sample space and concentrating on the induced sample space and the distribution of the random variable. From a practical perspective, random variables are either *discrete*, taking values in a finite or countably infinite subset of the real numbers (such as the integers) or *continuous*, taking values throughout the real numbers or subintervals. Any application can be modelled to be a mixture of both types, which are separated for analysis. In both cases the distribution is already uniquely defined by an easier concept, namely a point function.

Discrete case. Suppose a random variable X represents a 'count' and therefore takes non-negative integer values only. The sample space is the set $S_X = \{0,1,2,...\}$. An arbitrary assignment of the distribution is found by choosing any infinite sequence p_0, p_1, p_2,... of non-negative numbers adding up to 1, and then defining a probability for each 'singleton' event $\{i\}$ by the point function

$$i \to \mathbf{P}(\{i\}) := p_i \quad \text{for } i = 0,1,2,..$$

This function is called the discrete density or the probability density function of X. It uniquely determines the probability distribution P_X for any subset A of S_X, due to the additivity axiom A_3,

$$\mathbf{P}(A) = \sum_{i \in A} p_i$$

Intuitively, a total probability 'mass' of 1 unit is distributed among non-negative integers. A probability model in which a random variable can take any integer value, however large, is often useful, even in cases where there is an upper bound which the values cannot exceed.

Continuous case. Suppose now X represents a measurement on a continuous scale, taking values in all the real numbers. The general method of assignment is to take the total probability 'mass' of 1 unit and distribute it smoothly according to a *probability density function* $f_X(x)$, which is any non-negative function satisfying

$$\int_{-\infty}^{\infty} f_X(x)dx = 1$$

Its interpretation is that the probability of X lying in any interval $[a,b]$ is given by

$$P_X(a \le X \le b) = \int_a^b f_X(x)dx$$

The fact that this yields a legitimate assignment of probability follows from the additivity property of the integral operator and the mathematical property of probability measures being uniquely determined by their values on intervals. Since the integral is zero if $a=b$, any single value $X=x$ has zero probability.

If X is a random variable then its *mean* or expectation $E(X)$ is defined as the average value it can take, where 'averaging' is interpreted by weighing according to the distribution which is

$$E(X) = \sum_{i \in S_X} i\, p_X(i)$$

in the discrete case, and

$$E(X) = \int_{-\infty}^{\infty} x\, f_X(x)dx$$

in the continuous case. It may not be true that the above sums or integrals converge absolutely but if they do, we say that $E(X)$ exists since it is then finite. If $\mu = E(X)$ does exist, then it is a descriptive measure of the centre of a probability distribution like the arithmetic mean for a frequency distribution. It is also a guiding figure for the outcome of a series of experiments, just as probability can be linked to frequencies. If X is payment in a game of fortune then $E(X)$ is a guide to the payment one might expect - it is thus naturally called the fair prize of the game. Probability and expectation are intimately interwoven; historically, expectation was actually the more fundamental concept. If X counts the relative frequency of an event A occurring in a series of independent trials then

$$E(X) = P(A)$$

A measure of the extent to which a distribution is dispersed about the mean is the expectation of $(X-\mu)^2$, the variable measuring the squared deviation of single results from the mean, if it exists. This notion is known as the *variance* of X,

$$\text{var}(X) = E(X-\mu)^2$$

The variance of random variables and its square root, the standard deviation σ, play the same descriptive role for a probability distribution as their counterparts in descriptive statistics. Moreover, the mean and variance are the most important characteristics of distributions for practical and theoretical reasons. They are the fundamental ingredients of central theorems.

Central theorems

The concept of independence extends naturally from events to random variables. Two random variables X_1 and X_2 are independent if all events which may be described in terms of X_1, i.e. $(X_1 \varepsilon B_1)$ are independent from all events which may be described by X_2, i.e. $(X_2 \varepsilon B_2)$, where B_i are 'measurable' subsets of the real line. The probability of their intersection can thus be calculated by applying the simple multiplication rule

$$P[(X_1 \in B_1) \cap (X_2 \in B_2)] = P(X_1 \in B_1)\, P(X_2 \in B_2)$$

This equation can be used as a basis for defining stochastic independence. The notion of independence of a sequence of random variables involves no conceptual problem. A sequence of independent random variables X_1, X_2,... each with the same distribution is called a *random sample* from that distribution because it is a useful model for repeated observations. A single measurement of a quantity which is subject to random variation is represented by the distribution of the population. The measurement repeated under essentially the same conditions is represented by this sequence of random variables. The notion of a random sample is easily represented by a spinner which is independently spun several times.

The most important theorems of probability concern the limiting behaviour of sums of independent and identically distributed random variables.

Bernoulli's law of large numbers. Let A be an event of an experiment with $P(A)=p$, and X_i be determined by an occurrence of A in independent repetitions, i.e.

$$X_i = \begin{cases} 1 & \text{if } A \text{ occurs at } i\text{th trial} \\ 0 & \text{if } A \text{ fails to occur} \end{cases}$$

and let $H_n = X_1+X_2+...+X_n$ be the absolute frequency of A in n trials, then for any real positive number ε

$$\lim_{n \to \infty} P\left(\left|\frac{H_n}{n} - p\right| \geq \varepsilon \right) = 0$$

The variables X_i are represented by spinners which have only two sectors marked with 0 and 1. The theorem states that the relative frequency of the spinner coming to rest in the sector marked 1 in a series is 'convergent' to the probability p of this sector. It is the modern form of Bernoulli's 'theorema aureum' which for the first time related abstract and vague probability to concrete relative frequencies and thus opened the door for a broad field of applications. A generalization of it deals with an even stricter type of convergence; other generalizations refer to the convergence of the mean of samples to the mean of the underlying population:

If X_1, X_2, ..., X_n are independent random variables from a common distribution with finite mean μ and variance σ^2 then, given $\varepsilon>0$

$$\lim_{n \to \infty} P\left(\left|\frac{X_1+X_2+...+X_n}{n} - \mu \right| \geq \varepsilon \right) = 0$$

The law of large numbers states that the mean of a sample will be close to the (unknown) mean of a distribution from which the sample was drawn, with a high probability provided that the sample size is sufficiently large and the selection process is random.

Laplace's central limit theorem. The variable H_n (the absolute frequency) in Bernoulli's theorem will deviate from the expected value np in n independent trials so it is a random variable with its own distribution, worthy of study. De Moivre considered a special case and Laplace found H_n to be approximately normal.

$$\lim_{n\to\infty} P\left(\frac{H_n - np}{\sqrt{np(1-p)}} < z\right) = \int_{-\infty}^{z} \varphi(t)\,dt$$

with $\varphi(t)$ being the probability density function of the standard normal distribution.

The implicit condition of independence was sometimes overlooked when this distribution became regarded as a universal Law. The central limit theorem is a natural basis to approximate the distribution of the mean from random samples in order to derive confidence intervals or statistical tests for the (unknown) mean. These are the usual procedures to generalize information which is available only for a small subset of a population onto the population as a whole. The problem of statistical inference can be represented by a spinner whose marking are unknown. This spinner represents the population. We obtain the random sample X_1, X_2, ..., X_n by spinning the wheel n times and noting the outcomes. This sample information is the basis for finding the mean of the population. The central limit theorem states that, regardless of the shape of the original distribution, the mean of samples will approximately follow a normal distribution.

Standard situations

We now show how applications and concepts may be structured by a few distributions and their inherent model situations.

Laplacean experiments. These are experiments where the equidistribution is plausible on the basis of a physical symmetry. Conventional representations are spinners with equal sectors or urns filled with balls. For teaching, such experiments are useful to illustrate numerical probabilities to calibrate the weighing of uncertainty. They are a reference point to compare a real phenomenon to its symmetric counterpart in order to model it. Beyond their capacity to represent a real situation they also allow for the simulation of inherent consequences. Individual experience might be misleading; a six might seem intuitively less probable than other faces of the die as it is a highly desirable result, or as we remember the times waiting for a six but not for other faces.

Repeated Laplacean experiments. A Laplacean experiment is repeated and the sum or the mean of the results is of interest. The situation is easily represented by drawing balls from an urn or by spinning a pointer repeatedly. A random sample is the independent repetition of the same experiment. The special case of an experiment with two outcomes is known as a Bernoulli series. The distribution of the sum of two dice was historically an outstanding problem. To solve it formally, counting methods had first to be developed. It was disputed for a long time if the order of outcomes should be taken into consideration, e.g. if (2,3) is different from (3,2) with dice.

The role of independence in constructing probabilities on combined experiments, especially repeated experiments is neatly represented by spinners or urns. The conceptual difference between the mean of a sample, the mean of the population, and the random variable consisting of the mean of the samples is facilitated by repeated Laplacean experiments. Its influence on the central theorems may be studied by simulation; frequencies are only valid statistically if they stem from independent trials; furthermore, their reliability depends on the sample size.

Waiting time. With an experiment, it is of interest to ask for the waiting time for a specified outcome to occur. In a Bernoulli series with probability p the expected waiting time is $1/p$. Historically, expectation played a more fundamental role than probability in resolving famous problems. In evaluating individual probabilities, waiting time information stored in our memory is a fundamental ingredient. Unfortunately, this information is often misleading as some outcomes are relevant for ongoing events and others are of minor importance. Thus, waiting time experiences are different for the various outcomes and this may intuitively change the previous symmetry of the possible outcomes leading to biased probability judgements. Furthermore, the time to wait for an event which has not occurred, is independent of the already elapsed waiting time.

Poisson process. Poisson experiments may simply be introduced as Bernoulli series in which the number of trials is high and the probability p is small. An example is counting the atoms decaying in a specific period of time out of 1 kp of uran U_{238}. In this case the Poisson distribution is a mere tool for approximating binomial probabilities as no new concepts are involved. The Poisson process, however, describes a genuine random phenomenon of 'producing' events within time; heuristically, the process has to obey the following rules.

It does not matter when the observation of the process actually starts, the probability of various counts of events depend only on the length of observation. For short periods the probability to have exactly one event is essentially the intensity τ of the process to produce events multiplied by the length of observation. Furthermore for short periods one may neglect the probability of two or more events. Finally, events occur independently in time. The variable X which counts events in one unit of time then follows a Poisson distribution:

$$P(X=k) = \frac{\tau^k}{k!} e^{-\tau}, \qquad k = 0,1,2,...; \ \tau > 0$$

Many time-processes can be modelled by a Poisson distribution like the number of customers arriving at any service point, or the number of breakdowns of technical equipment. Conceptually, the Poisson process generalizes the idea of a Bernoulli series; the events are still counted but it is not possible to note when a trial is done. For instance, in counting the number of breakdowns, it is not clear what is to be meant by a single trial which can lead to a breakdown or not. As with waiting time in Bernoulli series there is a tendency for biased storing of facts in our memory. Furthermore, the evaluation of coincidences of events, the well known law of series, or special patterns in a random sequence might lead to misleading pre-concepts requiring discussion. The method of simulation can simplify the calculations.

Elementary errors. A normal experiment may simply be introduced as a Bernoulli series in which the number of trials is high and p is of intermediate value. The normal distribution is, by the central limit theorem, a good approximation for calculating binomial probabilities, at least if $np(1\text{-}p) > 9$. The same theorem, however, yields a formal representation of a genuine random phenomenon which can be modelled by the normal distribution. Historically, the hypothesis of elementary errors was formulated to explain why the error X of repeated measurements is normally distributed. This hypothesis was soon transferred to biometric variables and, for a long time, every quantity 'became normally distributed'. It still yields an intuitive interpretation of the normal model. An observed quantity X can be thought of as the result of many elementary influences or errors, i.e., $X = X_1+X_2+...+X_n$ and if the X_i are independent random variables, identically distributed, then the distribution of X is approximately normal.

The normal distribution is very important, both as a distribution describing observed quantities and as a fundamental assumption for classical procedures like regression or analysis of variance. Furthermore, the central limit theorem provides a sound basis for approximating the distribution of the sample mean $\overline{X}$ as normal and thus yields an easy calculation of confidence intervals and tests for the mean of a distribution.

3. Paradoxes and Fallacies

Though mathematics aims to deal with universal truths, its progress has not always been smooth. The history of mathematics has revealed many interesting paradoxes some of which have served as triggers for great change: probability is especially rich in this respect, as hinted at in earlier parts of this chapter. In a recent book, Székely (1986) shows how the theory developed from paradoxes and emphasizes the contradictions that have done most to clear up fundamental crises in the mathematics of randomness. He defines a paradox as a true but startling theorem, while a fallacy is a false result obtained by reasoning that seems correct. Each paradox is discussed in five parts: the history, formulation, explanation, remarks and references. As Székely states, the paradoxes and fallacies are both interesting and instructive. It is an exciting and novel approach; we give a flavour of its possibilities within this section, which will also be a useful reference in later chapters on psychological and pedagogical aspects.

Of course, it would be helpful to have a clear classification, but this is no mean achievement in that matters are often interwoven and overlapping. In addition, there is no general consensus about the underlying difficulties nor about how to best explain and overcome the traps. Ultimately, a classification comes down to personal taste. There are many paradoxes and fallacies so it is not possible to give a full or comprehensive listing. Instead, a few important paradoxes, representative of the different types, are discussed in detail. They are presented in four groups: chance assignment, expectation, independence and dependence, and logical curiosities. In each case, we formulate the paradox, give solutions and discuss underlying ideas.

Chance assignment

Laplace's definition of probability has several inherent difficulties. First, the 'favourable/ possible' rule has to be applied to an appropriate sample space which may differ from the space of all really possible cases as the 'Division of Stakes' shows. Fine's problem with the two dice has been dealt with in section 1 on history. It reveals that the relations between symmetry, equiprobability and independence are intuitively difficult as respecting ordering of the dice establishes independence and equiprobability on the full product space, whilst neglecting ordering destroys this independence. Second, the reduction of randomness to equiprobability of possible cases leads to situations where various models of possible cases result in conflicting probabilities. The different distributions in Fine's dice problem are an indicator; Bertrand's chord and the simpler library problem are further typical examples. The 'principle of insufficient reason' which is designed to classify equiprobable cases can thus lead to contradictions. Furthermore, this principle seems to turn complete ignorance into knowledge of equiprobabilities, which has already been discussed earlier.

Division of stakes. The problem has already been stated in section 1. At the score of 4:3 for A against B and 5 wins required for finishing the series, two more rounds will finish the game. The following four continuations of the game are fictitious but equiprobable:

5:3, 6:3; 4:4, 5:4; 5:3, 5:4; 4:4, 4:5.

Three lead to a final win of A, whence P(A wins) = 3/4 and P(B wins) = 1/4. In order to apply equiprobability via symmetry, one has to include the possibilities 5:3, 6:3 and 5:3, 5:4 which would never occur in practice as 5:3 stops the series. Thus, to apply the favourable to possible rule one has to include all *possible* cases, even those which cannot occur in reality. This hypothetical space is tricky and hence counterintuitive. This fallacy highlights the intuitive difficulty of enumerating cases which are not actually possible.

Bertrand's chord. An equilateral triangle is drawn in a circle with radius R and a line randomly drawn through the circle. What is the probability that the segment s of the line in the circle is longer than the side a of the triangle? Among the various solutions three are given here.

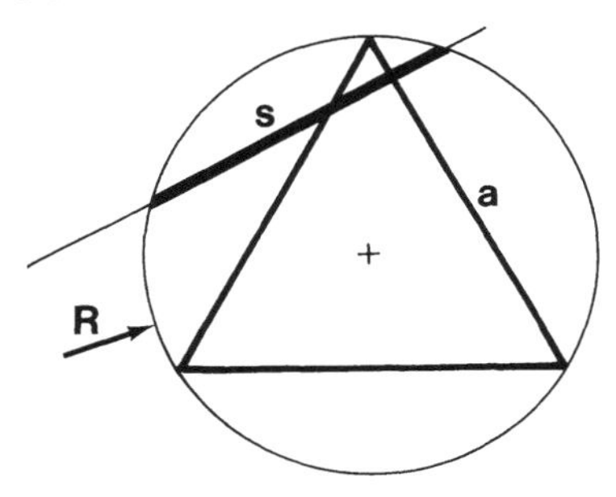

Fig.1: Line segment s and side a of the triangle to be compared

a) As the segment is uniquely determined by its mid-point M, we may focus our attention on the position of M. If M is contained in the circle with radius R_1 with $R_1=R/2$, we have $s>a$, otherwise $s\leq a$ (fig.2a). Hence

$$P(s>a) = \text{area}(R_1)/\text{area}(\text{circle}) = 1/4$$

b) We may compare the position of s with the diameter d, perpendicular to s. If s falls within the interval I with $|I|=R$, its length is greater than the length of a (fig.2b). Thus

$$P(s>a) = 1/2$$

c) As each segment s cuts the circle in P and Q we may consider the angle β between s and the tangent t at Q in order to express the position of P in terms of β, which can lie in range (0, 180). If $60<\beta<120$ we have $s>a$ (fig.2c), thus

$$P(s>a) = 1/3$$

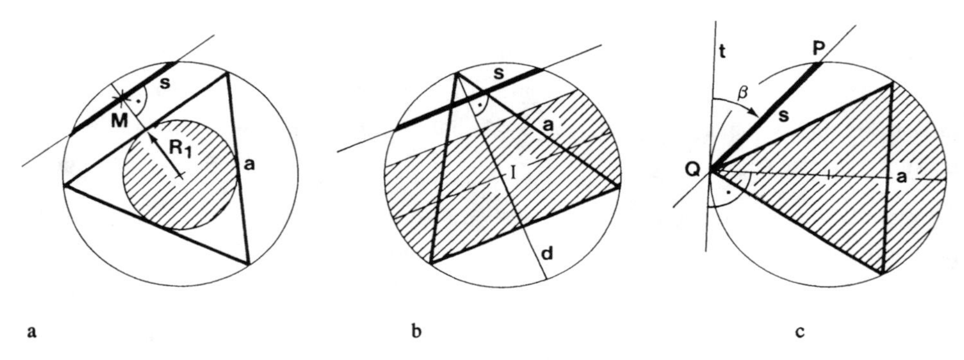

Fig.2a-c: Different possibilities to draw a random line

If chance is determined by the equiprobability of Laplace's definition, there should only be one set of possible cases and a unique probability. However, each of the three solutions represents chance and inherent equiprobability via its particular random generator. Is there more than one way of determining a line? This reflects an intuitive conflict and yields a contradiction to the basic assumption of Laplace's definition; the word *randomly* is neither fully covered by this approach nor is it meaningful without reference to an actual generator of the events.

There has been debate on whether there is a convincing argument for one of the above solutions which could rescue Laplace's position. If we require the natural assumption that translations and rotations of the plane do not affect the probability of the position of the line, then there is a unique answer; this nontrivial result is discussed in Palm (1983).

Library problem. Pick a book randomly from the university library. What is the probability that it is written in English?

a) Let E denote the number of English books and T the total number of books in the library. If these numbers are known, the required probability is E/T. The selection is easily managed by a random sample from an index. For E=500 and T=1000 the answer is 0.5.

b) For a different solution, take the procedure of selection as a starting point. Suppose that the library has a big and a small room, with a broad corridor in between where tossing a coin decides which room to enter. Moreover, suppose that the English books are not evenly distributed over the two rooms; for the numbers E_1=410, T_1=900, E_2=90, T_2=100, a short calculation yields the answer 0.678.

Of course, the selection procedure could be more complicated if a choice between the shelves is necessary etc. Again, if picking randomly is to be covered by Laplace's approach then all the possible ways of selecting a book should yield the same probability. But different selection processes clearly lead to different answers so the approach fails. Either the definition does not serve its purpose or it leads to differing results as in Bertrand's problem.

Expectation

The historical background of de Méré's problem reveals that expectation has been an alternative to probability long before either idea was formally defined; the St. Petersburg paradox and the example of dependent coins both illustrate the difficulties of expectation and its relationship to probability. Moreover, expectation is a key concept being the link between probability and statistics.

St. Petersburg paradox. Two players A and B toss one coin until it shows a 'head' for the first time. If this occurs at the nth trial, then player B pays £2^n to player A. What amount should A pay to B before this game starts in order to make it fair?

If X denotes the amount B pays, then its sample space is (in theory) all natural numbers. The expected gain is $\mathrm{E}(X)$, but this does not exist as the series involved diverges:

$$\mathrm{E}(X) = 2\cdot\frac{1}{2} + 4\cdot\frac{1}{4} + \ldots + 2^n\cdot\frac{1}{2^n} + \ldots$$

Thus player A would have to pay an infinite amount of money to B before the game starts.

Huygens introduced the expected value as the fair prize of a stochastic game. In this example, the fair prize is above all limits. The chance of a long sequence is very small and tends to zero, yet the expected payout is still infinite. So no one would or could play such a game, as the prize won is actually a limited amount of money. Another consideration is the expected length of the game, which is two rounds, a very short duration giving a normal payoff of $2^2=4$! These two values of 4 and infinity are counterintuitive and show that the relationship between different expectations is complex.

Dependent coins. A bag contains seven coins, their values are 100p, 50p, 50p, 50p, 10p, 10p, 10p. Three coins are drawn at random. What is the expected value of their sum? Is it relevant if the drawn coins are replaced?

The expectation of the first coin drawn is given by

$$E(X_1) = \frac{100 + 3 \cdot 50 + 3 \cdot 10}{7} = 40p$$

If coins are replaced, this is the same for the second and third draw, hence $E(X_2) = E(X_3) = 40p$. If the drawn coins are not replaced, then one has to use conditional expectations. The required expectation for the second draw is now the weighted mean of the three conditional expectations:

$$E(X_2) = \frac{1}{7}.E(X_2|100) + \frac{3}{7}.E(X_2|50) + \frac{3}{7}.E(X_2|10) = \frac{1}{7}.\frac{180}{6} + \frac{3}{7}.\frac{230}{6} + \frac{3}{7}.\frac{270}{6}$$

Calculations are tedious for the third draw but give the same result. Thus, the answer is 120p as the expected value for the three coins, whether or not coins are replaced between draws. In fact, this illustrates a fundamental and rather surprising property of expectation, its linearity. Whether random variables are independent or dependent,

$$E(X_1+X_2+X_3) = E(X_1) + E(X_2) + E(X_3)$$

provided the single expectations exist and are finite. For complications resulting from single expectations not being finite see Székely (1986, p.193).

This paradox characterizes the difference between probability and expectation. From the perspective of probability, the simplifying relation of linearity seems intuitively unacceptable if random variables are dependent. This seems contrary to the changes in probability calculations necessitated by non-replacement. However, for expectation there is an underlying symmetry; the probability of a specific coin being drawn at the first draw is the same as for the second or third try, even if coins are not replaced. Thus its individual contribution to *expected value* $E(X_1)$ is the same as for $E(X_2)$ and $E(X_3)$. As this is true for all coins,

$$E(X_1) = E(X_2) = E(X_3) = 40p$$

even if coins are *not* replaced and the coins which are drawn are not known. Thus the required expectation is 120p. There are straightforward mathematical arguments to support this intuitive reasoning. The paradox shows that intuitions about expectation need to be tackled in teaching which tends to focus on the probability concept.

Independence and dependence

Independence is a key concept in structural probability from the objectivistic perspective as it allows for the construction of models for repeated experiments, which are at the core of application. Dependence and conditional probability play a comparable key role within the subjectivist framework, via Bayes' formula. Inadequate intuitions abound: d'Alembert's paradox cannot be solved without the independence concept; Bertrand's cabinet shows how information affects the uniform distribution; Father Smith and son shows that the procedure of obtaining information actually influences probability.

D'Alembert's problem. Two coins are flipped. What is the probability that they yield different results?

a) The full combinatorial product space of *HH*, *HT*, *TH*, *TT* where order counts has a uniform distribution which yields the answer 1/2.

b) D'Alembert argued that the mixture *HT* has the same likelihood as the outcomes *HH*, *TT*, therefore the answer is 1/3. According to Bayes' rule of succession, the probability of *H* increases to 2/3 if the first trial is *H*; the detailed results are:

$$P(HH) = P(H)\,P(H|H) = \frac{1}{2}\cdot\frac{2}{3} = \frac{1}{3} \qquad P(TT) = P(T)\,P(T|T) = \frac{1}{2}\cdot\frac{2}{3} = \frac{1}{3}$$

$$P(HT) = P(H)\,P(T|H) = \frac{1}{2}\cdot\frac{1}{3} = \frac{1}{6} \qquad P(TH) = P(T)\,P(H|T) = \frac{1}{2}\cdot\frac{1}{3} = \frac{1}{6}$$

This might have been a strong justification for d'Alembert not to differentiate results according to the order of their appearance, hindering references to independence. D'Alembert's argument is theoretically consistent but neglects the huge amount of information available on coins so that one single toss would not lead to a change from 1/2 to 2/3 for heads or for tails. The underlying ideas need careful discussion in the classroom.

Bertrand's cabinet. A cabinet has three boxes each with two divisions. Three gold and three silver coins are put in the divisions so that two boxes contain coins of the same kind and one the mixture. Choose a box and open one of its divisions; it is assumed that it contains a gold coin. What is the probability of finding a gold coin in the other division?

a) The three boxes are equally probable, this feature is not changed by the choice of one; the information that there is a gold coin in the open division means that the pure silver box is no longer possible; the required probability is 1/2.

b) A formal though easy application of Bayes' formula yields the solution of 2/3.

It is a deep-seated fallacy that the given information about a gold coin will not change the equidistribution on boxes except that the probability should be applied to the reduced sample space of the two remaining boxes; there is also a reluctance to accept the results of Bayes' formula. There are various didactical strategies to overcome this problem. Freudenthal (1973, p.532) and Bentz (1983) use the technique of 'hidden lotteries'; the lottery on the boxes is symmetric, but the choice of the subdivision is, by no means, symmetric. Falk (1983) and Borovcnik (1987) suggest the favour concept which could intuitively clarify the higher estimate of the probability of the pure gold box as gold in the open division is circumstantial evidence for the box with two gold coins.

Father Smith and son. Mr Smith is known to have two children. He is seen in town and introduces the boy with him as his son. What is the probability that his other child is also male?

a) As boys and girls are (roughly) equally likely and births are independent, the information about one child is irrelevant, thus $p=1/2$.
b) The four possible combinations are BB, BG, GB, GG. The given information rules out the last possibility, thus $p=1/3$.
c) If BB or GG is true, there is no choice for Mr Smith. Suppose that he chooses a boy to accompany him with the probability q in case of BG or GB, then Bayes' formula yields

$$\text{P(second child a boy} \mid \text{we see Mr Smith with a boy)} = \frac{1}{1+2q}$$

If Mr Smith has no preference in the case of BG or GB, i.e. he chooses the child to accompany him with probability $q=1/2$, then the answer is $p=1/2$.

The strong intuitive drive towards b) or the difficulty of strengthening the vague argument of a) makes learners reject the formal solution in c) which could give insight if accepted: the independence of gender at birth is negated by preference (dependence) of companionship! Again, the hidden lotteries or the favour concept helps to clarify the situation intuitively.

Logical curiosities

The formal structure of probability is delineated by the axiomatic approach but the axioms do not regulate the structure of conclusions. These conclusions have a structure which differs from logical reasoning and causes various paradoxes. For instance, probabilistic conclusions lack transitivity, which is highly counterintuitive.

Intransitive spinners. Suppose there are three spinners. Which is the best to choose if two players compete?

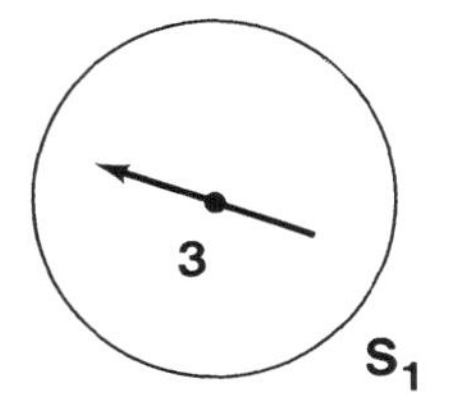

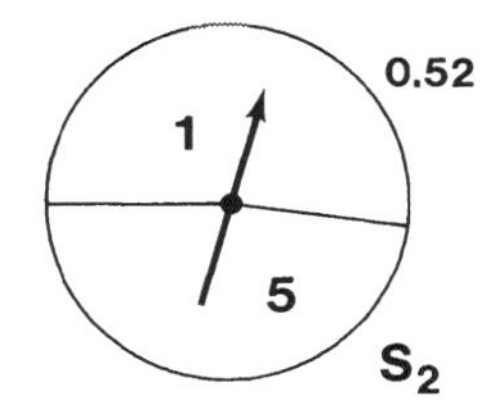

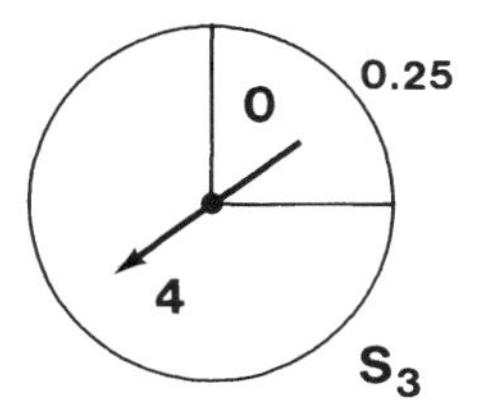

Fig.3: Intransitive spinners

As P(S_1>S_2) = 0.52, P(S_2>S_3) = 0.61, and P(S_1>S_3) = 0.25, the spinner S_1 beats S_2, S_2 beats S_3 yet S_1 does not beat S_3. There is no transitivity in the choice, any spinner beats and is beaten by one other spinner so the person who chooses second has an advantage.

This is a paradox from the perspective of ordinary logic where the transitivity of conclusions is normal. Stochastics is therefore not a weak form of logic, as expressed by saying '... is true with probability p' instead of '... is definitely true', but a different one. The favour concept serves as a method of reinterpreting such phenomena and clarifying the different structure of probabilistic reasoning.

Blythe's paradox. With the three spinners in fig.4 consider the following game: if each of two players has to choose a spinner, which one gives the highest number? Is the choice affected if a third player enters the game?

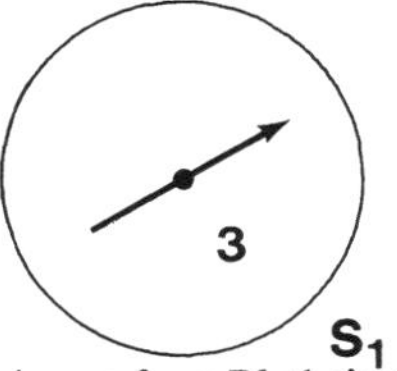

0.51
2
4
6
0.29
0.20
S_2

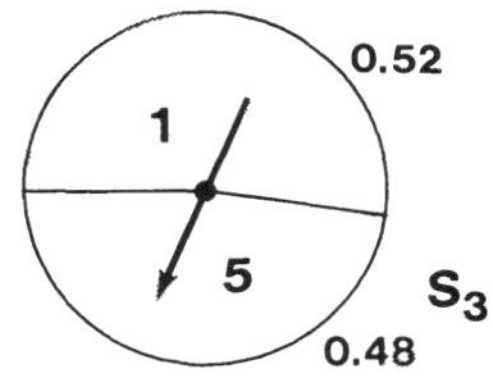

Fig.4: Spinners from Blythe's paradox

A short calculation yields (with two significant digits)

$$\begin{aligned} P(S_1 \text{ beats } S_2) &= 0.51 \\ P(S_1 \text{ beats } S_3) &= 0.52 \\ P(S_2 \text{ beats } S_3) &= 0.62 \end{aligned}$$

Thus S_1 is the best choice, the best second choice is S_2, no doubt. Now with a third player, a slightly longer calculation shows that S_3 is the best choice then:

$$\begin{aligned} P(S_1 \text{ beats } S_2 \textit{ AND } S_1 \text{ beats } S_3) &= 0.27 \\ P(S_2 \text{ beats } S_1 \textit{ AND } S_2 \text{ beats } S_3) &= 0.35 \\ P(S_3 \text{ beats } S_1 \textit{ AND } S_3 \text{ beats } S_2) &= 0.38 \end{aligned}$$

The result is surprising and intuitively unacceptable. The worst choice in a two person game is actually the best if three persons take part. The spinners are physically completely independent, yet the stochastical outcomes depend on each other in the sense of comparative probabilities. This is not obvious but is reflected by the relevant algebraic terms used in calculations. For example

$$P(S_3 \text{ beats } S_1 \textit{ AND } S_3 \text{ beats } S_2) \neq P(S_3 \text{ beats } S_1)\, P(S_3 \text{ beats } S_2)$$

Thus the multiplication rule does not hold which means the games are not stochastically independent.

In a variation of Reinhardt (1981) the spinners are placed as different tracks on the same spinner. It is obvious that in a two person game the outer track is favoured to win. In a three person game the inner track is best.

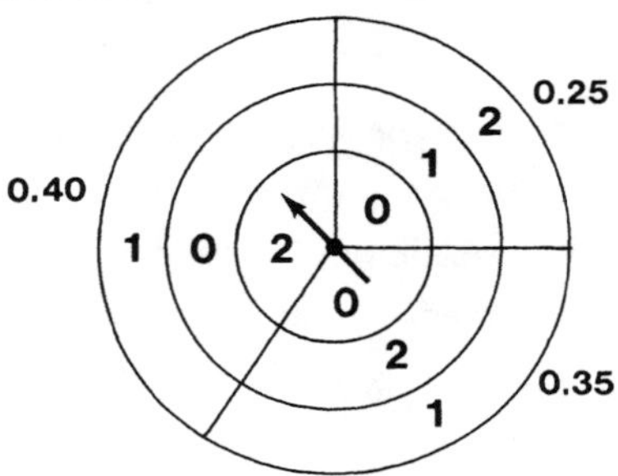

Fig.5: Reinhardt's spinners

The games have become physically dependent now, but that does not affect the stochastics. Varying the probabilities, given as fractions of the circumference, one can obtain any sort of game one wishes (in the sense of the paradox). Similar puzzles show that stochastical information cannot be inferred from the physical setting of the random variables involved. There is of course a formal way to overcome this: checking the multiplication rule.

Simpson's paradox. In 1973 the admission rate at the University of California at Berkeley of female applicants of 35% was lower than that of male applicants of 44%. Searching for the reason for this sexual 'discrimination', it turned out that in some departments women actually had higher admission rates than men while most of the departments had similar rates for both. Is it possible that admission rates for *all* departments are higher for women than for men, and nevertheless still lower for the whole university?

We simplify matters by assuming there are only two departments. Black marbles represent admission, white marbles rejection, so, in table 1 for example, 2 out of 5 females and 1 out of 3 males are accepted by department 1.

	Females		Males	
	Black	White	Black	White
Department 1	2	3	1	2
Department 2	3	1	5	2
University	5	4	6	4

Table 1: Illustrative data for Simpson's paradox

In both departments the proportion of black marbles (admitted) is higher for females than for males, 2/5>1/3 and 3/4>5/7. But for the university as a whole, the reverse holds: 5/9 for females admitted is lower than 6/10 for males. Again hidden lotteries give intuitive insight into this paradox. The reason lies in the different application rates of males and females; females tend to apply to departments with low admission rates. The paradox arises since the result seems contrary to ordinary logic, where dealing with separate cases is a valuable method of proof. If case i) and ii) cover all the possibilities and are mutually exclusive then a relation is usually proved if it is shown to hold both in case i) and ii). This structural feature, however, does not hold for probabilistic reasoning.

Concluding comments

Paradoxes and fallacies can be entertaining. They can raise class interest and motivation. Discussion of these ideas can help to

- analyze obscure or complex probabilistic situations properly
- understand the basic concepts in this field better
- interpret formulations and results more effectively
- educate probabilistic intuition and reasoning firmly.

The multitude of misconceptions in the examples shows that probabilistic intuitions seem to be one of the poorest among our natural and developed senses. Perhaps, this is a reflection of the desire for deterministic explanation. We have great difficulty in grasping the origins and effects of chance and randomness: we search for pattern and order even amongst chaos. The examples above illustrate the gap between intuition and mathematical theory, particularly because stochastic reasoning has no empirical control to revise inadequate strategies. Paradoxes and fallacies highlight these difficulties as they are signs of a cognitive conflict between an intuitive level of reasoning and formalized, mathematical arguments. In a paradox, the 'objective' side is adequate but intuitively inaccessible, whereas in a fallacy the objective side is inadequate though intuitively straightforward. Thus, planned discussion can foster individual conceptual progress.

Bibliography

d'Alembert, J.: 1754, 'Croie ou pile', in *Encyclopédie*, vol.4, Paris.

Arbuthnot, J.: 1710, 'An Argument for Divine Providence Taken from the Constant Regularity Observed in the Birth of Both Sexes', *Philosophical Transactions of the Royal Society* **27**, 186-190.

Barnett, V.: 1973, *Comparative Statistical Inference*, Wiley, New York.

Bayes, T.: 1763, 'An Essay towards Solving a Problem in the Doctrine of Chances', *Philosophical Transactions of the Royal Society* **53**, 370-418; reprinted in E.S. Pearson and M.G. Kendall: 1970, *Studies in the History of Statistics and Probability* , vol.1, Griffin, London, 131-154.

Bentz, H.-J.(ed.) 1983: *Probleme im Umgang mit dem Zufall, Der Mathematik-Unterricht* **29**(1).

Bernoulli, D.: 1777, 1961, 'The Most Probable Choice between Several Discrepant Observations and the Formation therefrom of the Most Likely Induction', *Acta Acad. Petropolitanae*, Petersburg, 3-23; English translation *Biometrika* **48**, 3-18.

Bernoulli, J.: 1713, *Ars conjectandi*, Basel.

Bertrand, J.: 1899, *Calcul des Probabilités*, Gauthier-Villars, Paris.

Bessel, F.W.: 1818, 'Untersuchungen über die Wahrscheinlichkeit der Beobachtungsfehler', *Astronomische Nachrichten* **15**, 358-359, 368-404.

Bickel, P.J., E.A. Hammel, and W.J. O'Conell: 1977, 'Sex Bias in Graduate Admissions: Data from Berkeley', in W.B. Fairley and F.M. Mosteller (eds.): 1977, *Statistics and Public Policy*, Addison-Wesley, Reading, Mass.

Blythe, C.R.: 1972, 'Some Probability Paradoxes in Choice from among Random Alternatives', *Journal of the American Statistical Association* **67**, 366-381.

Borovcnik, M.: 1987, 'Revising Probabilities according to New Information - A Fundamental Stochastic Intuition', in *Proc. of the Sec. Int. Conf. on Teach. Stat.*, Victoria, 298-302.

Buffon, G.L.: 1785, *Natural History, General and Particular*, 2nd ed., Strahan and Cadell, London.

Cardano G.: approx. 1560, 'Liber de ludo aleae', published 1663 in *Opera omnia*, vol.1.

Chebyshev, P.L.: 1887, French transl. 1890-91, 'Sur Deux Théoremes Relatifs aux Probabilité', *Acta arithmetica* **XIV**, 805-815.

David, F.N.: 1962, *Games, Gods and Gambling*, Griffin, London.

Edwards, A.W.F.: 1978, 'Commentary on the Arguments of Thomas Bayes', *Scandinavian Journal of Statistics* **5**, 116-118.

Falk, R. and M. Bar-Hillel: 1983, 'Probabilistic Dependence between Events', *Two-Year-College-Mathematics Journal* **14**, 240-247.

Feller, W.: 1957, *An Introduction to Probability Theory and Its Applications* 2nd ed., Wiley, New York.

Fermat, P. de and B. Pascal: 1654, *Correspondence*, publ. 1679 in Toulouse; English translation in F.N. David: 1962, *Games, Gods and Gambling*, 229-253, Griffin, London.

Fine, T.L.: 1973, *Theories of Probability*, Academic Press, New York.

Finetti, B. de: 1937, 'La Prévision: Ses Lois Logiques, Ses Sources Subjectives', *Ann. de l'Inst. H. Poincaré* **7**, English transl. in H.E. Kyburg and H.E. Smokler (eds.): 1964, *Studies in Subjective Probability*, Wiley, New York, 93-158.

Freudenthal, H.: 1973, *Mathematik als pädagogische Aufgabe*, vol.2, Klett, Stuttgart.

Galton, F.: 1889, *Natural Inheritance*, Macmillan, London.

Gauss, C.F.: 1809, *Theoria motus corporum coelestium in sectionibus conicis solem ambientium*, Perthes und Besser, Hamburg.

Graunt, J.: 1662, 'Natural and Political Observations upon the Bills of Mortality, Chiefly with Reference to the Government, Religion, Trade, Growth, Air, Diseases etc. of the City of London', *Royal Society of London.*

Huygens, C.: 1657, 'De ratiociniis in ludo aleae', in F. v. Schooten: *Exercitationes matematicae*, Leyden.

Jeffreys, H.: 1933, *Theory of Probability*, Oxford University Press, London and New York; 3rd ed. 1969.

Kendall, M.G.: 1956, 'The Beginnings of a Probability Calculus', *Biometrika* 43, 1-14, reprinted in E.S. Pearson and M.G. Kendall: 1970, *Studies in the History of Probability*, vol.I, Griffin, London, 19-34.

Kolmogorov, A.N.: 1933, *Grundbegriffe der Wahrscheinlichkeitsrechnung*, Springer, Berlin; reprinted 1977.

Kolmogorov, A.N: 1956, *Foundations of the Theory of Probability*, Chelsea, London.

Laplace, P.S. de: 1812, 'Essay Philosophique sur les Probabilités', *Journal de l'École Polytechnique* **VII/VIII**, 140-172, English translation 1951, Dover, New York.

Laplace, P.S. de: 1814, *Théorie Analytique des Probabilités*, 2nd ed., Courcier, Paris.

Maistrov, L.E.: 1974, *Probability Theory. A Historical Sketch*, Academic Press, New York.

Markov, A.A.: 1913, 'A Probabilistic Limit Theorem for the Cases of Academician A.M. Lyapunov', repr. in A.A. Markov: 1951, *Selected Works of A.A. Markov in the Theory of Numbers and the Theory of Probability, Akad. Nauk*, 319-338.

Mises, R. v.: 1919, 'Grundlagen der Wahrscheinlichkeitsrechnung', *Mathematische Zeitschrift* **5**, 52-99.

Moivre, A. de: 1738, *The Doctrine of Chances*, 2nd ed., London.

Ore, O.: 1953, *Cardano: The Gambling Scholar*, Princeton University Press, Princeton, New Jersey.

Palm, G.: 1983, 'Wo kommen die Wahrscheinlichkeiten eigentlich her?', *Der Mathematik-Unterricht* **29**(1), 50-61.

Pacioli, L.: 1494, *Summa de arithmetica, geometria, proportioni et proportionalitá*, Venedig.

Pearson, K.: 1902, 'On the Systematic Fitting of Curves to Observations and Measurements - Part I', *Biometrika* **1**, 265-303.

Peverone, G.F.: 1558, *Due brevi e facili trattati, il primo d'arithmetica, l'altro di geometria.*

Poisson, S.D.: 1837, *Recherches sur la Probabilité des Judgements en Matiére Criminelle et en Matiére Civile*, Bachelier, Paris.

Quetelet, A.: 1835, *Sur l'Homme et le Dévelopement de Ses Facultés, ou Essai de Physique Sociale*, Paris.

Reinhardt, H.E.: 1981, 'Some Statistical Paradoxes', in A.P. Shulte and J.R. Smart, *Teaching Statistics and Probability*, National Council of Teachers of Mathematics, Reston, 100-108.

Schneider, I.: 1989, *Die Entwicklung der Wahrscheinlichkeitstheorie von den Anfängen bis 1933 - Einführung und Texte*, Akademie-Verlag, Berlin.

Schnorr, C.P.: 1972, *Zufälligkeit und Wahrscheinlichkeit*, Berlin.

Simpson, T.: 1755, 'A Letter to the President of the Royal Society, on the Advantage of Taking the Mean of a Number of Observations in Practical Astronomy', *Philosophical Transactions of the Royal Society* A **49**, 82-93.

Stegmüller, W.: 1973, *Probleme und Resultate der Wissenschaftstheorie und Analytischen Philosophie*, vol.4, 1st part: Personelle Wahrscheinlichkeit und Rationale Entscheidung, 2nd part: Personelle und statistische Wahrscheinlichkeit, Springer, Berlin-New York.

Székely, G.J.: 1986, *Paradoxes in Probability and Mathematical Statistics*, Reidel, Dordrecht/Boston.

Manfred Borovcnik
Institut für Mathematik
Universität Klagenfurt
Sterneckstraße 15
A-9020 Klagenfurt
Austria

Hans-Joachim Bentz
Fachbereich Mathematik/Informatik
Universität Hildesheim
Samelsonplatz 1
DW-3200 Hildesheim
FR. Germany

Ramesh Kapadia
9 Beechwood Close
Surbiton
Surrey KT6 6PF
U.K.

Chapter 3

Empirical Research in Understanding Probability

M. Borovcnik and H.-J. Bentz

The analysis of historical development and philosophical ideas has shown the multifaceted character of probability. Kolmogorov's axiomatic structure does not reflect the complexity of ideas. The abundance of paradoxes not only occurred in the historical development of the discipline, it is also apparent in the individual learning process. Misconceptions are obstacles to comprehending and accepting theoretical ideas. Empirical research on probabilistic thinking aims to clarify and classify such misconceptions from both the theoretical as well as the individual's perspective. We present major research ideas of psychology and didactics of mathematics from a critical perspective. Our method of interpreting subjects' responses to experimental situations will be a complementarity of intuitions and official mathematics which is especially helpful for transferring ideas to actual teaching.

1. Research Framework

Fischbein's interplay between intuitions and mathematics forms the background for our reinterpretation of empirical research on probability concepts. This dynamic view serves as a key to interpret the communication between interviewer and subjects; it is also useful to describe the process of learning a theory. Empirical research is designed to elicit as far as possible the vague world of preconceptions. To initiate the communication, the subject has to be confronted with images and problems related to theory. Any further impetus at the interviewer's side may be regarded as partial instruction in theory which alters the status of prior intuitions. This is not a hindrance in teaching where such a change is intentional, but may cloud the existing preconceptions. Furthermore, subjects can communicate only their secondary intuitions which emerge from mathematics. To talk about these intuitions, they have to use parts of the theory itself, perhaps chains of corresponding formal signs or related pictures. Of course, if the interplay between the

R. Kapadia and M. Borovcnik (eds.), Chance Encounters: Probability in Education, 73–105.

interviewer's theoretical demands and subjects' intuitions breaks down, the process may become meaningless.

Peculiarities of stochastics and its teaching

Primary intuitions related to the concept of probability are often vague and confused: 'Probably it will rain tomorrow', 'The series 1,2,3,4,5,6 has no chance of occurring in the state lottery' etc. It is difficult to operate with such shaky intuitions. This is very different from the notion of natural numbers. Children count and calculate with quantities in their everyday life. Furthermore, a check of success is easy with natural numbers where the calculation of 2+2 yields the immediately testable result of 4. Such feedback is complicated with probability; the extremely small probability of winning in a state lottery is countervailed by the fact that people win every week.

The axiomatic foundation of probability yields a justification for the mathematical existence of the concepts and a regulation for their use *within* mathematics, i.e., how to combine probabilities if one already has values for them. The axioms give no clue as to which probability values to assign to events. In contrast to other theories, for probability there are no immediately evident actions, which could sharpen and revise conjectures. The intuitive comprehension of a theory and the secondary intuitions are not controlled by a mechanism independent of the theory and is therefore not reliable. Without understanding the whole theory, one cannot see how mathematics changes primary intuitions. This is true even for numbers, but it is not relevant until high levels of abstraction when, for example, the rational numbers prove to be countable which is clearly counterintuitive as the rational numbers have no obvious ordering like the natural numbers.

The controversy in the foundations between subjectivists (e.g. de Finetti) and objectivists (e.g. Fisher) reflects different types of primary intuitions of probability; they use their theoretical conception to justify their intuitions. This controversy is thought to be unresolved so that scientists now try to expose the parts in common between the various schools. Since the different axioms lead to equivalent theories, the Kolmogorov approach may be viewed as the unique theory of probability. But the different primary intuitions which are combined into one and the same theory lead to different secondary intuitions; these influence the conceptions of *statistics* which are still different in the various positions today.

The common name stochastics for the separate fields of probability and statistics obscures these differences. However, it is not possible to reduce the multitude of competing intuitions by reference to the formal equality of the system of axioms on the part of *probability* theory. Vice versa, the system of axioms of Kolmogorov alone, separated from intuitive conceptions, does not allow for a unique reconstruction of intuitions.

The role of paradoxes signifies a further peculiarity of probability; a multitude of puzzles accompanied the history of this discipline which did not occur to the same extent in other disciplines. These paradoxes either stem from a conflict between different intuitive ideas or they mark points where intuitive ideas and mathematics diverge. It is this feature of paradoxes which constitutes their educational value.

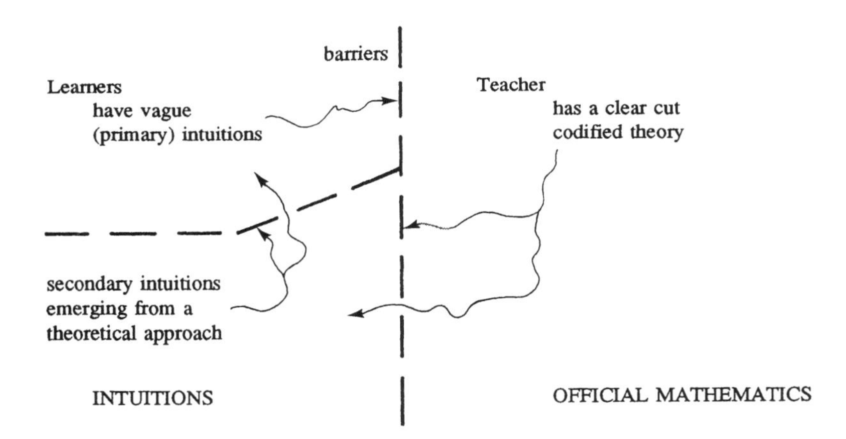

Fig.1: Sparse relations between intuitions and official mathematics

We illustrate some crucial points of teaching probability by this interplay of intuitions. Fig.1 shows that conventional teaching establishes too few links between primary intuitions and the mathematical model. This is critical for probability as there are no direct experiences which help learners to establish these links on their own. But, if learners are to understand mathematics and apply it to other situations, teaching has to start from the learners' intuitions and to change them gradually. The learner needs to see how mathematics will reconstruct his/her intuitions.

Research in psychology and didactics

Scientific research on individuals' intuitive conceptions has a long tradition within psychology and the didactics of probability. We name but a few investigations: Piaget and

Inhelder (1951), Cohen and Hansel (1956), Kahneman and Tversky (1982), Fischbein (1975), Falk (1983), Green (1982), APU-study (1983), Hawkins and Kapadia (1984).

Psychologists investigate cognitive processes in actual thinking. They describe types of probabilistic thinking and design theories to study its development in the course of maturation of individuals. These thinking processes are analyzed essentially as being independent from mathematical concepts. Note that these notions are not separable against the background of the complementarity of intuitions and mathematics. Thus, for example, the current version of mathematical concepts represents a series of intuitions.

Instead of paradigms of thinking itself, didactics of mathematics focuses on questions of intuitive acceptance of theoretical concepts and related ideas. The starting points are different probability theories and the underlying mathematical concepts. Pertinent questions are 'How do people understand, accept, or use certain mathematical concepts?', 'Which misconceptions are prevalent?', 'How should teaching be designed?' etc. There is a much closer link between cognitive processes and the means to develop them.

We will critically analyze empirical research and draw conclusions for teaching probability. Our intentions in this chapter are similar to Kapadia (1984, p.1):

> "The aim ... is to synthesize ... research undertaken to investigate how children develop their ideas of probability and the associated difficulties they encounter. It is intended to look at work done both in pedagogical and psychological areas, as well as considering mathematical aspects. It is hoped that this would be valuable in the classroom in offering the basis to construct coherent teaching approaches. ... An important feature is that selected, major work will be appraised, in a critical light."

Researchers have designed experimental situations to investigate if subjects are ready to respect and apply certain concepts. The overall assumption seems to be that the items represent a situation which allows for one intuition and a *unique* solution so that certain answers and the reasoning illustrate the understanding of concepts. We believe that this approach is flawed. Even the coin tossing context has many degrees of freedom; for example, a coin may be asymmetric, or real magic or cheating may be operative. Depending on the subject's perception of the situation, he/she may reduce the degrees of freedom differently and end up with a different but appropriate solution. In much empirical research, however, the world of intuitive conceptions has usually been perceived narrowly, both with regard to mathematics and to subjects so that interpretations are sometimes dubious.

We will describe this interplay between intuitions and mathematics by reconstructing the context of items in different ways. This will reveal pitfalls where the communication between interviewer and subject, or in the classroom, might break down. We also explore typical categories of thinking which characterize probability and how they are related to the world of the individual's intuitions.

Items of similar type occur in different investigations; we deliberately chose one prototype. The items are grouped according to four main problem fields: sample space and symmetry, frequentist interpretation, independence and dependence, and statistical inference. We reconstruct the items and trace strategies which lead to answers different from those which were expected and accepted by the researcher. The experiments are presented briefly with an interpretation of the empirical finding of the author; under the heading 'problems' we show the complexity of an item by developing this scenario of multiple analysis; we also link our findings to teaching.

We do not aim at a critique of the general framework of the research. However, we elaborate phenomena to show the variety of conceptions in the minds of subjects and researchers - sometimes causing a breakdown in communication. We want to demonstrate how difficult these investigations are if reliable conclusions are to be drawn (and we really do aim at this!) and, where the difficulties arise. Furthermore, we want to illustrate that communication in class is similar to the experimental situation and hence our analysis might improve actual teaching, too.

2. Sample Space and Symmetry View

Symmetry played a key role in the history of probability. Pascal and Fermat's famous correspondence (1654) revealed that they both had a good awareness of the set of possible outcomes of a random experiment even if they did not refer to an explicit definition of probability based on such a reference set. Initially, de Méré found that his experience in actual games did not coincide with theory; his difficulty was the lack of a concept of sample space. It was only Laplace (1812) who clarified the application of the favourable/possible rule.

Laplace's attempt to define probability is characterized by an intimate intermixture of sample space (the mathematical part) and the idea of symmetry (the intuitive part). A probability is fixed by equal likelihood of all possible results. In modern mathematics the sample space is completely separated from the probability which is a function defined on a specific class of subsets of the sample space. But, the concept of sample space cannot fully be understood if it is not related to intuitions of symmetry.

Teachers want children to develop certain fundamental concepts from mathematics such as probability distribution based on a sample space. Expectation, however, might be a more promising initial concept. Therefore, it is of interest to study whether individuals are willing to accept underlying images of equiprobability and sample space, and under which circumstances and at which developmental stages they do so. We discuss two items by Green (1982) from his study of about 3000 pupils aged 11-16 years.

No.1: Tossing a counter

Item. A small round counter is *red* on one side and *green* on the other side. It is held with the *red* face up and tossed high in the air. It spins and then lands. Which side is more likely to be face up, or is there no difference? Tick the correct answer.

(*R*) The *red* side is more likely.
(*G*) The *green* side is more likely.
(=) There is no difference.
(*?*) Don't know.

Result. (*R*)	(*G*)	(=)*	(*?*)
18↓11	28↓12	45↑74	5 %

The answer considered to be correct is marked by an asterisk*. The notation 45↑74 means that 45% of 11 year olds and 74% of 16 year olds chose answer (=). If there is no significant variation related to age, we report the mean over all years 11-16. The figures do not total 100% as they represent averages over different age groups.

Interpretation. The percentage of correct answers for this simple question is relatively small. This suggests that children aged 11-16 have a rather wayward intuition of probability. One response (*G*) is expected because of the formulation of the problem. Answering (*G*) is induced by the so called 'negative recency effect'. By the age of 16, symmetry prevails over other views.

Problems. We discuss various potential strategies to reveal the complexity of the item and of interpreting the answers. We do not claim that they are all in the minds of the children.

0) The normative strategy: This is the strategy to yield the usually accepted solution by standard methods. After having stripped the problem of all irrelevant information one 'sees' the sample space $S=\{G,R\}$. Symmetry considerations yield a uniform distribution which gives the answer (=).

1) Verbal understanding: The subject looks at the pattern of the underlined words: 'red - green - red' which may lead to answer (*G*). Second, the chain of associations 'red face up' plus 'it spins' plus 'it spins one time' leads to answer (*G*). Third, the subject vividly imagines the toss, as this yields a specific result, answering (*R*), (*G*), or perhaps (*?*). The first two approaches might be induced by suggestive elements of the text like underlining or the vivid description of 'it spins'; the third by the description of a concrete trial but also by a general perception of the problem as it is more useful to predict the concrete result.

2) Causal thinking: We sketch the interference with causal thinking by reference to *coin tossing*. If coin tossing is a physical experiment then the same conditions of performing the experiment would result in the same fall each time. It seems promising to study the physics of coin tossing in order to predict the concrete result under specific conditions. However, the normative solution implies neglecting the physics with the exception of the symmetry of the coin; coin tossing is used as a paradigmatic image for probability of 1/2, instead. This probability is an unpromising statement as it gives no clue to the concrete result of a trial. Causality may be suggested by the wording of the item. Any answer could be expected then.
Another form of causality is related to the inability to predict specific results. The coin can land on either 'head' or 'tail'. Both results are likewise unpredictable in the sense that there is no guessing procedure that guarantees 100% success; therefore the answer (=) is chosen. This naive causal symmetry would also attribute equal weights if the process of abstraction in 0) yielded considerably different weights.

3) Reflection-decision conflict: The child's reflection ends up with a 50:50 estimate, i.e. $P(G)=P(R)=1/2$. But which of the answers to choose if the child is forced to do so? The result of reflection does not help for the actual decision; we see a discrepancy between concept and strategy. One of the answers (*R*), (*G*), (*?*) is expected. If a subject really is in this conflict and is asked to justify the answer, he/she may be confused because there is no justification available. This conflict is even more likely to occur if the item is more orientated toward action which is

expressed by the wording 'which result do you expect?' or 'which outcome would you bet on?'

It is unlikely that the pupils use the normative strategy to end up with the 'correct' answer. Unfortunately the answers offer no help to discriminate between this and other strategies. It is striking that this simple item was mastered only by 45% of eleven year olds. This performance is also confirmed by our own research with students of mathematics. Is this due to the suggestive elements which may have favoured interference with causal thinking or is this due to the reflection-decision conflict? We do not feel it appropriate to assume "fallacious thinking" (Green, 1982, p.14).

Teaching. This item is particularly appropriate to the idea of probability. Probability does not allow for predicting the result of a specific trial even if there were enough computer power at one's disposal. Instead, by means of probability only a subsidiary problem is solved. The solution is in terms of probability; the probability of 1/2 for 'head' is a very indirect type of knowledge for the next trial.

Probabilistic justifications are peculiar. If $P(G)=5/6$ and $P(R)=1/6$, this is an intuitively understandable justification for the choice of G for the next trial but $P(G)=P(R)=1/2$ justifies the choice of G as well as of R. How can a justification for an event at the same time justify its negation?

Interference with causal thinking is deep-seated. Probabilistic experiments are replicable under the same conditions like experiments in science. Nevertheless they show very distinct features. If the conditions are perfectly reproduced, results will still vary. Confusion between probabilistic and causal thinking may be decreased by a careful discussion.

There is a conflict between the concept as a result of reflection and problem solving strategies as a decision orientated device to find a concrete solution. It is different to deliberate on abstract weights for the possibilities or to choose one possibility as a result for the next trial. The cognitive processes involved are distinct and lead to different results. If one does not pay attention to the actual mode of cognitive processes within students, a breakdown in communication in class might follow. Our own investigations show that verbal understanding is shaky, changing from one moment to the next; slight changes in phrasing the item may have unforeseen consequences. Furthermore, extrinsic information such as text layout has a sizeable effect.

No.2: Hat lottery

Item. A mathematics class has 13 boys and 16 girls in it. Each pupil's name is written on a slip of paper. All the slips are put in a hat. The teacher picks out one slip without looking. Tick the correct sentence:

(*B*) The name is more likely to be a *boy* than a girl.
(*G*) The name is more likely to be a *girl* than a boy.
(=) It is just as likely to be a girl as a boy.
(*?*) Don't know.

Result.	(*B*)	(*G*)*	(=)	(*?*)	
	4	38↑71	53↓25	2	%

Interpretation. The low achievement of 16 year olds of 71% is remarkable though this item is slightly more difficult than experiment no.1 as there are symmetries on single children or on categories of boys and girls. Kapadia and Ibott (1987) report on a gender difference as 81% of boys but only 61% of girls were 'correct'. This implies that the pseudo-reality of the context led the girls to use their own common sense notions from reality.

Problems. Again we discuss potential strategies for dealing with the problem.

0) Normative strategy: Symmetry on $S=\{g_1,...,g_{16},b_1,...,b_{13}\}$ yields (*G*).

1) Further symmetries: A symmetry strategy on $S^*=\{B,G\}$ yields (=).

2) Pseudo-reality of context: The context of the item makes other information vivid which might induce subjects to use information from the item only if it seems relevant. The numbers 13 and 16 are neglected intentionally as they yield proportions of 0.45 and 0.55 which do not differ so much from the prior 1/2. Or, the pseudo-real features might induce the subject to use other problem solving strategies or a different item reconstruction like 'there is roughly the same number of girls and boys in this world' (without recognizing the numbers 13 and 16), or 'boys are preferred in this world', or 'there are more boys in my class'. Such reconstructions lead to an unconscious neglect of relevant information.

3) Causal thinking: Either a girl or a boy is drawn. For both events it is completely impossible to predict with certainty the actual result. 'Boy' is as unpredictable as 'girl'. This symmetric ignorance leads to answer (=).

4) Reflection-decision conflict: The conclusion '50:50' or 'girl is more probable' (reflection) and the assumption that only one symbol will be written on the slip actually drawn (action) yields (*B*), (*G*) or (*?*). A subsidiary non probabilistic strategy may determine the actual choice so that asking for reasons might induce a breakdown in communication.

The methods used to solve the problem may be classified in two categories. One refers to the *process of abstraction* which recognizes the extent to which the experiment is symmetric, i.e., which sample space is to be chosen; the other category refers to numerical *skills* in determining the solution which includes the use of fractions.

The process of abstraction is predominant in the examples discussed here, the numbers involved are simple. There is a series of five experiments in Green's investigation related to the comparison of chances of two given urns. The situation in these items is highly pre-structured so the abstraction process required is trivial. Thus the main source of problem lies in the numerical skills. If numbers involved are also simple, the percentage of correct answers is up to 90% even at the age of 11. Compare this with the poor achievement on experiments no.1 and 2.

The discussion reveals that even with simple experiments, judging the difficulty of the abstraction process required is an enormously hard task for the experimenter. We gave an inkling of the variety of other influences which play a role in an individual's approach to an item. So it is difficult to link the level of problem solving strategies to the level of related concepts. The difficulties increase if one has to rely on summary statistics which give no clue to the individual's perception of the item. We believe that judgements of the complexity should be done carefully *before* any interview. This would improve the pertinent discussion significantly.

Teaching. We discuss some further issues. Urn problems have been criticized for their artificiality long before mathematics didactics came into being. To meet this critique, real examples have been introduced into teaching. These examples, however, are only pseudo-real and may be no better than their artificial counterparts but are simply different. They should be used in teaching but their peculiarities still need attention.

Artificial context items are usually drawn from abstract theory and may serve as links between intuitions and theory. However, they do not coincide with the related abstract concepts. Someone with no experience of them may intuitively make adequate use of the underlying concept in other, more familiar situations. Artificial items can serve as a medium between abstract concepts and intuitions only if subjects have sufficient experience, perhaps via simulation.

Pseudo-real context items must be described in terms of an artificial context (or an abstract model codified in symbols which is no easier) before they can be solved analytically. The problem with pseudo-real items is thus twofold. First, the learners have to master the corresponding artificial situation which may not be within the reach of their 'intuitive world'. Second, they may not be able to describe the item by means of its artificial counterpart as they have their own experience from the pseudo-real context. Teaching pseudo-real items can conflict with common sense, a poor basis for teaching.

There is another problem resulting from the specific type of knowledge provided by probability statements. Random experiments are usually explained to be experiments in which one lacks any knowledge to predict the specific result. The complete inability at the intuitive level, however, is often modelled by equal chances at the theory level: to have no sufficient reason to prefer one of the results is Laplace's principle of applying equal likelihoods to all possibilities. A subject might choose the answer of equal probability despite the differing numbers for boys and girls and justify this by an argument like "as it is luck" or "it can be either" as is reported by Kapadia and Ibbott (1987). One cannot equate complete ignorance in the sense of not being able to predict specific outcomes with the symmetry from the Laplace principle; teaching has to clarify whether this ignorance is simply an intuitive expression for the *probabilistic* feature of a situation.

3. Frequentist Interpretation

How does randomness manifest itself in repeated trials? Limiting frequencies are theoretically covered by the law of large numbers. The many misconceptions underline the complexity of the concept of the limit of relative frequencies. Another aspect refers to the patterns of specific occurrences. There is an intuitive reliance on logical, causal or even magical patterns of outcomes. In the face of five consecutive heads (H), a change to tail (T) seems long overdue. These patterns may also have been an obstacle to derive the concept of sample space and the related symmetry view which is a *static* view on just one trial.

The mathematics of random patterns is complicated. A sequence of length n is one element of the combinatorial product space, e.g., tossing a coin six times with the special results of *HHH HHH* or *HHT THT*. The probability distribution on this product space is derived by means of the concept of independence of single trials. (We will discuss difficulties related to independence later on.) If coin tossing obeys the hypothesis of independence, both sequences above are random and have the same probability. If this assumption is doubtful, a statistical test is applied.

Most people lack experience in conscious analysis of sequences which in due course could change primary intuitions. In judging for randomness, individuals tend to rely on a set of primary intuitions distinct from the usual probabilistic concepts. The pattern of a particular sequence itself is often believed, by its logical or causal features, to reveal whether or not it is random. These logical or causal features may also emerge from inadequate secondary intuitions of the concept of sample space. Within psychological research there is an endeavour to identify concepts and cognitive strategies which form and guide thinking. Kahneman and Tversky (1972, 1982) have described intuitive strategies by just three heuristics, namely representativeness based on similarity or distance in connotation, availability based on ease of retrieval or associative distance, and causal schemes based on establishing causal chains.

It is of interest to study the primary intuitions which guide a subject's thinking on problems related to random patterns and the secondary intuitions emerging from probability concepts which lead to such heuristic strategies. Furthermore, to what extent can individual behaviour be described by means of such heuristics and are they stable within individuals? We discuss one item by Shaughnessy (1983) related to the representativeness heuristics and one more from Green (1982) which is a spatial analogue.

No.3: The six children

Item. Shaughnessy (1981) posed 70 college students (in an introductory course in probability) the following question thereby varying an experiment by Kahneman and Tversky (1972).

The probability of having a baby boy is about 1/2. Which of these sequences is more likely to occur for having six children?

(A) *BGGBGB* *(B)* *BGGBGB* *(C)* *about the same chance for each*

Result.	*(A)*	*(B)*	*(C)**	
	70	3	27	%

Interpretation. The sequence (*A*) seems to be more representative for the process of births because the relative frequency of boys equals the probability of boys. Many people seem to base their judgements on such similarities. "Their responses often indicated that they felt this sequence fits more closely with the 50:50 expected ratio of boys to girls" (Shaughnessy, 1981, p.91)

Problems.

0) Normative strategy: Symmetry arguments on $S=\{B,G\}$ yield equiprobability for boys and girls, i.e., $P(B)=P(G)=1/2$. Due to the independence assumption between the gender of different children all elements of the product space $SxSxSxSxSxS$ have the same probability $(1/2)^6$:

$P(BGGBGB) = P(B)P(G)P(G)P(B)P(G)P(B) = (1/2)^6 \approx 0.015$.
Thus answer (*C*) is correct.

1) Verbal understanding: The answer (*C*) 'same chance for each' may have been confused with a probability 1/2 for each sequence. This conflicts with the correct intuition that the outcomes (*A*) and (*B*) are highly improbable; each sequence of length 6 has the small probability of 0.015. This might confuse a subject completely.
Another source of confusion lies in the wording. There is an easy transition from the statement 'one special sequence with three boys' to '*one* sequence ...'; losing the accent on 'one' results in 'one sequence ...' which is very similar to 'one arbitrary sequence...'. These subtleties may prompt subjects to identify sequence (*A*) with the proposition 'no. of boys equals 3', and (*B*) with '..equals 5'. Of course, P(3 boys) > P(5 boys). If subjects base their answers on classes of sequences, i.e. the number of boys, and not on single sequences, they are right.

2) Pseudo-reality of context: Of course, the probability of a boy is not exactly 1/2; furthermore, there seems to be evidence for a slight dependence of gender within a family. Yet subjects are expected to assume independence as in coin tossing. Common sense may hinder students from establishing this correspondence; or, deep-seated emotions may restrain them from modelling choosing a child by tossing a coin. In real problems there is often no interest in the special ordering of the sequence but only in the numbers. The resulting transfer or the obstacles above thus may induce subjects to neglect order so that they end up with solution (*A*). Our own investigations indicate that subjects are more ready to respect ordering in the coin tossing equivalent of item 3.

3) Reflection-decision conflict: Even though a probability is required in this item, the subjects tend to perceive it in a decision orientated way which is closer to their intuitions. Our own investigations show that subjects favour answer (*A*) to an even greater extent with a decision orientated item.

Teaching. A new aspect of this item relates to verbal understanding. The language of probability often conflicts with intuitive ideas. 'Same chance' is not the same as 'probability 1/2', since it also means 'same degree of improbability'. Furthermore, slight changes of wording prompt a very different problem reconstruction such as single sequences rather than classes of sequences. There is a need for carefully designed teaching to obtain feedback on actual problem reconstruction. This is not only necessary for the correct solution of individual problems but also for developing adequate concepts.

Normative reconstruction conflicts with common sense as there is an urge to neglect ordering; this reconstruction is reinforced by the pseudo-real features of item 3. Pseudo-reality thus may be an obstacle to successfully integrating more realistic problems in teaching.

The reflection-decision conflict may also be operative in item 3. Whenever theoretical considerations lead to equiprobable results, they offer no guide on which possibility to choose. The subject has to rely on other strategies to make the required choice if the problem is posed in a decision orientated way or if it is reformulated in such a way as individuals tend to do. As these may not be sensible, a breakdown in communication is likely.

Scholz (see chapter 7) criticizes the attempt to explain subjects' behaviour by a few heuristics because no cognitive framework is developed which allows insight into when and how a specific heuristic is applied by an individual. From psychological literature it seems that such heuristics only account for errors in probability problems. They could however, be a fundamental intuitive key to understand theoretical ideas; Bentz and Borovcnik (1986) discuss the links between representativeness and random selection.

No.4: Snowfall

Item. The flat roof of a garden shed has 16 square sections. It begins to snow. At first just a few snowflakes fall, then after a while more have landed. Below are three sets of two pictures. Each set in fig.2 shows the pattern of snowflakes building up - first 4 flakes, then 16 flakes. Which of these sets best shows the kind of pattern you would expect to see as the snowflakes land? (We have made some slight change in the layout of

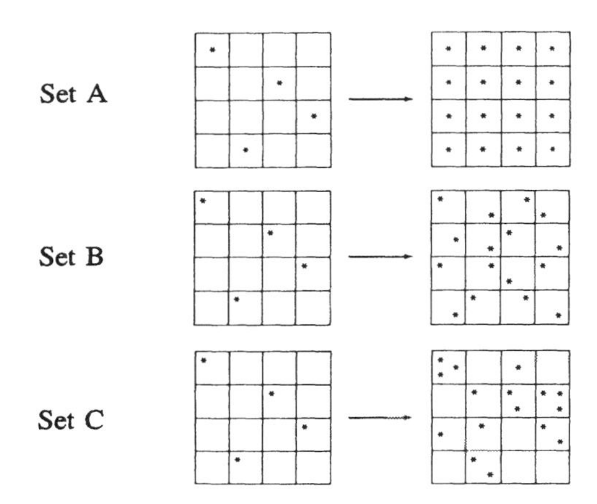

Fig.2: Patterns of snowflakes on the tiles of a roof

the possible answers.)

(*A*) (*B*) (*C*) (*BvC*) (*Each kind of pattern is as likely*)

Result.

(*A*)	(*B*)	(*C*)*	(*BvC*)	(=)	
12	14	26↓18	22	27↑34	%

Interpretation. Children do not have adequate nor clear cut ideas of effects of randomness. Green goes even beyond that in his interpretation of the fact that the rate of success decreases with age:

> "It can be hypothesized that we see at work two opposing tendencies - maturation/experience on the one hand and dominance of mathematical/scientific deductivism on the other which stifles the appreciation of randomness by seeking to codify and explain everything." (Green, 1983, p.774).

Problems.

0) Normative solution: Snowfall is modelled by a Poisson process. The number of snowflakes on one tile has thus a Poisson distribution; its parameter τ is estimated from the mean count of all 16 tiles: $\tau \approx 16/16 = 1$.

Number of snowflakes		0	1	2	3
probabilities		0.37	0.37	0.18	0.06
Observed	in *A*	0	1	0	
frequencies	in *B*	0	1	0	
	in *C*	7/16≈ 0.44	4/16≈ 0.25	3/16≈ 0.19	2/16≈ 0.12

Inspection by eye gives a good agreement between the Poisson distribution and the observed frequencies in set *C*; only this set passes a X^2-test which leads to answer (*C*). Of course this item was not intended to be solved by the Poisson model above. The intention might have been to ask for related primary intuitions. A simulation of 'snowing' with equal probabilities for each tile would show that there will be some tiles free of snow and others containing several snowflakes.

1) Further normative strategies: In the normative modelling it does not matter if snowflakes land in the middle or on the edge of tiles; it is only the frequency that counts. However, remembering the special (time) ordering of snowflakes, each of the sets is equally likely. Therefore, the answer (=) also fits into normative frame. If one asks for the expected number τ of snowflakes on each tile, then set *A* and *B* are the right answer as exactly one snowflake per tile is to be expected.

2) Verbal understanding: The graphical layout is highly confusing. In set *A* the snowflakes land in the middle of tiles, in set *B* and *C* they are dispersed on the tiles.
'Pattern' usually means 'regularity', an idea which may also be suggested by the pseudo-real features of the item - snow covers the surface evenly. This leads to answer (*A*). If one rejects this idea, what else could be meant by 'pattern'? Is it the individual formation of a pattern or is it the equivalence classes of similar patterns? If a child actually does refer to individual sequences, the answer 'equally likely' is correct.
The interpretation of 'how *could* this pattern develop' as 'how *should* ..' or a literal interpretation of 'would you expect' may provoke answers in terms of expected values, while 'best shows the kind of pattern' is relatively vague.

3) Pseudo-reality of item: People have many impressions about snowfall like 'snow falls evenly'. This might reinforce a certain type of verbal understanding (regularity) and lead to answer (*A*) or to answer (=). But children may be confused by the fact that they have no experience of snowfall at this micro level (16 snowflakes!) They might think hard on how this pattern of 16 snowflakes could or should develop. This might lead them astray.

To summarize, the problem is posed in a very vague way. There are 'reasonable' probabilistic conceptions to any of the answers which make them correct. Of course it is difficult for children to have a clear image of the various model conceptions and even more to verbalize what they think. It is, however, easy to imagine that they did hesitate in giving their answer because they had an awareness of the difficulties.

Children lack intuitions about the clustering of random events as needed in strategy 1). They have no relevant experience as the Poisson process describes only small parts of reality; situations in which the parameter τ does not vary with time or space are unrealistic. Simulated experience with the pure phenomenon of the Poisson process could supply this experience but is usually not available.

If subjects have not understood the problem and if an answer is forced (the possibility 'don't know' is not included in this item), many of them refer to symmetry considerations and regard all the possibilities they are offered as equally likely. It is particularly interesting that there is some learning effect for this strategy with age (27↑34). This incoherent strategy is a sign of a complete breakdown in communication; it indicates the complexity of the problem rather than children's inadequate conceptions of random patterns.

Teaching. The pseudo-real context and complexity of the problem with too many degrees of freedom may cause a breakdown in communication with children relying on subsidiary incoherent strategies. The complexity of intuitions related to a problem field is often reduced too quickly to an abstract mathematical model to which the learner associates nothing at the intuitive level. However, these intuitions, inadequate or adequate, have to be brought out if teaching is to develop children's thinking.

4. Independence and Dependence

Independence is one of the key concepts of probability theory. Probability (at least for the objectivists) is designed as a theoretical concept which builds upon the intuitive idea of relative frequencies in independent trials. Independence comes into the theory via a trivial definition which is usually justified by vague, almost mystic arguments. Thus the basis of (primary) intuitions is not clarified. On the other hand, the independence concept is the key assumption for central theorems which establish secondary intuitions on probability. From the perspective of the interplay between mathematics and intuitions, the definition of independence is far from trivial, it is a crucial point. Subjectivists avoid the difficulties by substituting it by exchangeability, an intuitively more accessible concept.

Dependence is not a simple complement to the notion of independence as there is not one dependence but a whole range. It is formalized by the unconditional and conditional probabilities of an event in question being different. In the objectivist approach, dependence is a derived concept; as such it should not be crucial for the interpretation of probability. The basic axioms, however, do not cover primary intuitions related to dependence nor do they supply secondary intuitions which help to change them. In fact, deep-seated intuitions emerging from causal thinking, conflict with the mathematical notion of dependence. The subjectivist approach of probability justifies it as degree of belief; conditional probability is a basic notion which covers the primary intuitive idea of revising probability judgements as new information becomes available. Thus the world of intuitions related to dependence is now crucial for a deeper understanding of probability.

It is remarkable that intuitions forming the opposing positions of subjectivists and objectivists are often unconsciously intermixed and may confuse the learners. It is of interest to note those tasks where the discrepancy emerges and intuitions obstruct an adequate reconstruction and its related solution. We will integrate some characteristics of independence in the subsection on statistical inference which is intimately interwoven with independence. Here, one experiment shows the difficulties of designing suitable items to investigate children's intuitions about independence, another reveals the intuitive problems subjects encounter with these key concepts.

No.5: Dependent urns

Item. The actual instructions given to the children were as follows. I have 12 beakers in a row. Each beaker contains a lot of beads, which may be blue or yellow, but the proportions of blue and yellow beads vary from beaker to beaker. I am going to take out four beads, one at a time, from each beaker. I will show you the first three beads as they are drawn and then you will guess the colour of the fourth (Cohen and Hansel, 1956, p.14).

The experiment was designed in such a way that in each arrangement of three beads drawn from a beaker there were always two of one colour and one of the other. Each of the six arrangements was repeated twice, the whole series of 12 being given in random order, as illustrated below: *YBB*, *YBY*, *BYB*, *BYY*, *BBY*, *BYB*, *YYB*, *BYY*, *BBY*, *YBB*, *YYB*, *YBY*.

Result. Cohen and Hansel (1956) interviewed 93 children aged 10+. Their data revealed that in 68% of predictions children incorrectly chose the non predominant colour (p.15). Indeed the non predominant colour was chosen more often the longer it did not occur in the series. With the three possibilities *YBB*, *BYB*, *BBY* for example, the prediction *Y* for the fourth bead is chosen most often in case of *YBB*, least often in the case of *BBY*. In item no.8 which is quite similar, 66% of the subjects prefer the answer 'equally likely' instead of the non predominant colour.

Interpretation. Many individuals find it difficult to consider an independent outcome isolated from a series of outcomes of similar type. Children show a tendency to choose the non predominant colour because they assume that the proportions of different colours in the beaker are the same. They still believe that the proportions are equal after one yellow and two blue beads are drawn so that the yellow beads are predominant in the urn then. In this respect it is 'more likely' that the fourth bead is yellow. Or, simply they may not like asymmetric arrangements, whereby the choice of the colour of the fourth bead gives them the opportunity to establish a balance.

Problems.

0) Normative solutions: To solve the problem, the subject has to make the assumption of $P(Y)=P(B)=1/2$ and independence for the process of filling the beakers. Filling the beakers should be like coin tossing. Drawing from one beaker subsequently is just a rearrangement of the original filling process which does not alter independence. Thus the already drawn beads yield no information on the colour of the next. This modelling yields *B* or *Y* for prediction. However, if the probabilities for yellow and blue are unknown, then the sample information has to be used to estimate the probability; in the case of *YBY* this yields an estimate of $P(Y)=2/3$ and the prediction of *Y* which is the dominant colour.

1) Effect from cheating: Though the task was intended to probe children's notions about independence, the actual experiment was designed in such a way that the outcomes *BBB* and *YYY* could not occur. So the experiment does not deal with an 'independent' process. Such 'cheating' could confuse the children completely or lead to a learning effect as the children had to make 12 ensuing predictions. The repeated non-occurrence of *YYY* or *BBB* indicates that there are no beakers with 4:0, 3:1, 1:3, or 0:4 'yellow:blue' composition. This makes 2:2 beakers more

probable and the prediction of the non-predominant colour more sensible from one trial to the next.

2) Symmetry and equalizing strategies: There are two types of symmetry, namely related to the filling process and to the drawn sample. The first symmetry does not help to predict the outcome of the fourth bead, whereas the other symmetry directly leads to an equalizing strategy and to the 'solution'.

3) Reflection-decision conflict: A child might have some notion of independence and would like to answer 'it does not matter'. As this possibility is not included in the catalogue of answers the subject is forced to choose either *Y* or *B*; in doing so he/she follows an equalizing strategy lacking better arguments for other strategies.

The item as a whole is complex. If children do not understand it, they have to rely on some subsidiary strategy such as equalizing the colours of drawn beads, which arises in the reflection-decision conflict. This is supported by the answer of one child to a related problem. "Because usually the one who has been losing a lot wins. But really it is just luck". (ibid., p.33 of appendix). The first sentence describes a strategy to find an answer but the second one reflects the appreciation of non-dependence of the outcome in question. We believe that there is a considerable difference between exploring the existence of a special formation of a *concept* in the imaginative world of the child and tracing a problem solving *strategy* within the range of its capabilities.

The effect of cheating is hard to allow for. We believe that some children did learn in the course of the experiment and some children were extremely confused by it. There is no way to differentiate between answers according to a positive learning effect and those according to an equalizing symmetry strategy, both favour the non predominant colour: in the case of *YBB*, e.g., both lead to prediction *Y*.

Teaching. The complexity of tasks is likely to be underestimated and may be increased by a cheating framework. The formation of concepts has to be based on appropriate reference situations. If tasks are complicated, the probability increases that learners simply do not understand and have to rely on subsidiary strategies. A child who is asked for a justification of a subsidiary strategy may give any answer, so no sensible interpretation is possible.

Whenever the item framework is actually decision orientated or is reconstructed to be decision orientated and the solution at the concept level yields no (obvious) hint for a decision then a conflict may be operating. The teacher will not understand the student if he/she expects an answer at a concept level and the subject actually gives it at a problem solving level.

No.6: Independent urns

Item. An urn contains two white balls and two black balls. We shake the urn thoroughly, and blindly draw two balls, one after the other, without replacement.

a) What is the probability that the second ball is white, given that the first ball is white?

b) What is the probability for the first ball being white, given that the second is white and the colour of the first is not known?

Result. Falk (1983) confronted her students at the university with this item; most of them had no difficulty in deriving 1/3 for a) while 50% responded with 1/2 for b).

Interpretation. Question a) causes no problems as there is a specific urn with three balls, two of them black and one white. Question b) is more subtle, leading half of the students to answer 1/2. The reason given is that "the first ball does not care whether the second ball was white or black." Since the second draw cannot affect the first, the required probability depends only on the urn's composition at the outset. The subjects' refusal to consider evidence occurring later than the judged event reflects their reasoning along the time axis. It is congruent with viewing probability as a measure of some feature of the objective urn situation, while the heart of the problem lies in its being addressed to our internal locus of uncertainty. One is asked here to revise one's current state of knowledge in the light of new evidence and to infer backwards.

Problems.

0) Normative solution: Drawing at each stage of the process is assumed to be equiprobable. From a tree diagram and the definition of conditional probability, the solution is 1/3; a formal approach is by means of product spaces.

1) Common sense: Some subjects find it simply stupid to hide the first ball and look at the colour of the second, before giving a statement on the first ball. That is not drawing balls from an urn for them. As this approach conflicts with their common sense they might be confused or use causal or time-bound thinking.

2) Causal thinking: First causality has to be established, then the effect will follow. At the beginning there is a 2:2 urn, the second ball cannot influence the first drawn so the probability is 1/2. Thinking along the time axis is closely related to

causal thinking but a more general approach. According to that, subjects simply refuse to judge a previous by a later event as this reverses the real course of action. The causal perception or the thinking along the time may lead to a confusion as subjects can see no reason why the second ball being white is given to them as extra information.

3) Symmetry: Out of the three unknown balls there are two black and one white ball which gives an answer of 1/3 for the first ball being white. It does not matter which of the white balls is still unknown. Subjects show a great reluctance to apply symmetry within this time dependent context.

Teaching. Tree diagrams are a useful tool for modelling situations. Like a bookkeeping algorithm, they allow modelling step by step and combining these steps to an overall result. This algorithm easily provides the answer but the approach camouflages the problem as there is no clue as to why deep-seated causal ideas or time dependent thinking are inadequate. If a mathematical method easily provides the solution but does not allow for any control over its mechanism at an intuitive level, then it is likely that students will not internalize the method but will continue to rely on their (inadequate) intuitions. Such teaching is doomed to fail. Formal methods are very powerful tools - it is precisely their inherent advantage not to illuminate their mechanism but at least in the initial phase the learner has to be convinced of their effectiveness at the intuitive level.

Urns are a standard reference point but may induce a causal connotation of probability. It is useful to have a suitable transformation of a problem to an urn from which numerical probabilities are easily read off. However, sometimes, this linking process is incorrectly generalized as learners might believe that probabilities make no sense without referring to an urn. Thus, probability shifts its meaning: an urn does not cause the specific outcome directly but does affect the probability value. Conditional probabilities reinforce this 'causal' interpretation as they are usually calculated in situations like question a) where at each stage the new conditional probability is represented by an urn.

'Causal' arguments are also used in motivating the multiplication rule for independent events when it is stressed that situations should be modelled by stochastic independence if they lack causal influence. This is a misleading link to the vague but tight interrelations between causal and stochastic independence. Experiment no.6 is an illuminating counterexample as 'white on the first draw' is causally independent of 'white on the second draw' but not stochastically!

We have tried variations of this experiment. We faced students with the situation that the second and third ball drawn are white and asked for the probability for the first to be white. Their immediate answer was 'zero'. However, in the case of an urn with 4 white and 4 black balls many of them still answer '1/2' giving causal justifications. Only in the situation that the second to the fifth ball was white, did they switch to 'zero'; after this it is logically necessary that a black ball must have been drawn at first.

Logical, causal, and probabilistic thinking are hierarchically ordered. Subjects seem to use causal independence and change their strategy only when logical reasons force it. They only acknowledged the value of the information about the second and subsequently drawn balls when we changed to an urn with 100 black and 100 white balls and presented the scenario of the second to hundredth drawn ball being white. Then, it was still not logically necessary for the first ball to be black, but the later events were recognized to be pertinent in evaluating the event 'white on the first draw'.

Experiment no.6 is valuable in illuminating independence and related ideas, and the different types of reasoning. It also helps to understand when information should lead to a new judgement of a situation which is formalized by a new conditional probability.

5. Statistical Inference

There are at least two conflicting intuitive ideas of probability which form the core of the positions of objectivists and subjectivists. Despite their different primary intuitions they have a common syntax. However, their secondary intuitions are different which leads to totally different statistical theories; this marks the heart of the controversy in the foundations of statistics. Probability theory is where the modelling of probabilistic phenomena occurs, while statistics regulates the application of probability to the real world. In the following experiments we will demonstrate the procedures of subjectivists and objectivists with regards to making inferences about unknown 'causes' and to predict the future.

The sources of problems for learners are at three levels. The type of question in statistics is complex and interferes with general and deep-seated primary intuitions like hope or fear. The methods in statistics are complex, hard to understand and far from primary and secondary intuitions in probability. Methods within the objectivist position conflict with primary intuitions which can be linked to the subjectivist conception and vice versa; the statistical test of a hypothesis, for example, conflicts with the primary intuitions about evaluating a hypothesis which is important for subjectivists.

What has been said about competing conceptions underlying the theory may only be speculated on with regards to the cognitive development of a child. It may be that a child's vague primary intuitions oscillate between the two positions during cognitive development. It is of interest to learn which conceptions children have, which reference situations could help to clarify the different levels of intuitions, and which intuitions are more helpful. We discuss two examples from Green (1982).

No.7: Coin tossing

Item. An ordinary coin is tossed *five times* and 'Heads' appears *every time*. Tick the correct sentence below:

(*H*) Next time the coin is more likely to turn up 'Heads' again.
(*T*) Next time the coin is more likely to turn up 'Tails'.
(=) Next time 'Heads' is as likely as 'Tails'.
(*?*) Don't know.

Result.	(*H*)	(*T*)	(=)*	
	11	12	67↑80	%

Interpretation. Green (1982) classifies this problem to be of probabilistic type which is different to our classification here. Green's conclusion about children's answers is of fallacious thinking but he gives no detailed interpretation in his summary paper. It has remarkable features in common with experiment 8 below and we will use it to develop typical statistical ways of thinking by elaborating the difference between these two experiments.

Problems.

0) Normative strategies: The hypothesis P(H)=1/2 and the sample information *HHH HH* are combined to give a new evaluation of P(H)=1/2. We speak of a hypothesis being 'strong' if there is strong evidence (or belief) in its being true in the illustrated situation. Clearly, P(H)=1/2 is a strong hypothesis for coin tossing; together with the independence assumption, it yields the answer (=). If the belief in this hypothesis is weaker, then a test is applied with α being the type I error. For α=1% this yields non rejection of P(H)=1/2 and the answer (=). A weaker

belief is reflected by a bigger value of α. For α=5%, P(H)=1/2 is rejected in favour of P(H)>1/2. This new hypothesis plus independence yields the answer (H).

1) Naive symmetry strategies: A naive symmetry relative to the sample space $\{H,T\}$ leads to (=). But symmetry relative to the actual sample leads to the equalizing answer (T).

2) Cheating: The biased information of five heads may induce subjects to rephrase a non probabilistic context for the item; cheating seems very plausible. Therefore they might be confused what to answer. In our own interviews, students asked for more information to exclude this possibility.

3) Reflection-decision conflict: If the child ends up with a 50:50 evaluation, the result of reflection gives no clue for the prediction of the sixth trial. However, the child might feel urged to choose a specific possibility (H) or (T) and rephrase the question in a decision orientated way as 'which face will show up next?'

4) Pattern recognition: Logical patterns in the sample shown may be used for prediction. Depending on the specific pattern, the answers (H) and (T) are possible. The logical pattern leading to (T) seems to be more attractive to many.

The phrasing for the outcome of the sixth trial is counterintuitive in itself as individuals experience fluctuation of coin tossing and their inability to predict concrete results. On the other hand, there is a deep-seated intuitive demand for a method which allows for such a prediction (see Konold, 1989). The formal statement of P(H)=1/2 may be neglected as it is very indirect information about the outcome of the sixth trial. Causal schemes, which are attractive elsewhere, are not available in this experiment, but logical reasoning about patterns seems to apply directly. If subjects use incoherent strategies, they might lack common sense justifications for the actual choice so that further arguments have to be interpreted with caution. It is difficult, perhaps impossible to refer specific answers to specific strategies, especially if the interviewer is given subsidiary justifications when communication breaks down. The methods involved in the normative strategy are highly complex and far from being intuitive.

Teaching. How an item is reconstructed by a subject is not trivial, it is crucial for any interpretation of an answer. Even in simple artificial contexts, the idea of cheating or an action orientated rephrasing might mislead. Teachers need a clear vision of several possible strategies and should not follow the one solution which is presumed to be normative. This simple experiment no.7 shows clearly that there are different normative strategies; if this is ignored there may be a breakdown in sensible communication. Besides, this experiment helps to clarify different types of probabilistic information. How should one distinguish between the information P(*H*)=1/2 on coins in general from this specific coin and the context in which it is thrown or the sample information *HHH*?

No.8: Drawing from a bag

Item. A bag has in it some white balls and some black balls. A boy [blindly] picks out a ball, notes its colour, and puts it back. He then shakes the bag. He does this *four* times. He picks a *black* ball *every time*. He then picks out one more ball. What colour do you think he is likely to get? Tick the correct sentence:

(*B*) Black is more likely again.
(=) Black and white are equally likely.
(*W*) White is more likely this time.

Result.	(*B*)*	(=)	(*W*)	
	13↑25	67	24↓10	%

Interpretation. "Drawing inferences from sampling experiments was another weak area, as judged by responses. Pupils tended to rely on the abstract idea of 'equality' and not relate it to the physical situation." (Green, 1982, p.15). "Items involving an appreciation of randomness, the stability of frequencies and inference were particularly poorly done, with little improvement with age evident." (Green, 1983, p.780).

Problems. We chose this item for two reasons. It is very similar to that from Cohen and Hansel (experiment no.5). Furthermore, it is very similar to coin tossing (experiment no.7) but it differs in important aspects. When we conducted our own interviews we recognized the potential of the interplay between these two items for teaching.

0) Normative strategies: As nothing is known for the distribution of balls and as P(*B*) is not known therefore, frequencies could be used. The hypothesis P(*B*)>1/2 is better supported than its complement, giving the answer (*B*). However, as P(*B*) is completely unknown it is more appropriate to summarize sample information by means of a confidence interval. A short calculation yields the 95%-interval (0.3, 1) which is much too imprecise. Thus, one should get more sample information before one answers the problem.

1) Additional information about the hypothesis: If there is a strong belief in the hypothesis P(*B*)=1/2, then one cannot learn anything from the sample; this view plus the independence assumption lead to (=). If the belief in this hypothesis is not so strong, a statistical test should be applied to include the sample informa-

tion. For α=5% the hypothesis is not rejected which in turn yields the answer (=). However, α=10% leads to a rejection of it in favour of P(*B*)>1/2 which justifies the answer (*B*). Furthermore, if one employs a subjectivist approach for the parameter P(*B*), e.g. a uniform distribution for P(*B*) over the interval (0, 1), and combines this with the sample information of four blacks then one obtains a probability of 5/6 for the next draw being black, favouring answer (*B*).

2) Verbal understanding: 'Some white balls and some black balls' may be equated to mean equal numbers of both colours, this leads to the strong hypothesis P(*B*)=1/2. Many subjects ignore that balls are put back as drawing without replacement makes more sense to them. Such a perception of the problem leads to the answer (*W*).

3) Naive symmetries: The sample space {*B*,*W*} yields (=). Looking at samples yields an equalizing answer, (*W*).

4) Reflection-decision conflict: The child ends up with the judgement of 50:50 for the chances for the next ball being black. But, what to do then if a decision is required? Either answer (*B*) or (*W*) can be expected.

The normative strategy fails in the context of the example because the sample is simply too small which is clearly shown by the confidence interval. One does not end up with a satisfactory answer unless one makes assumptions within the problem context; the solutions, however, depend on these assumptions which, in turn, depend heavily on the conception of probability regarded as appropriate.

Experiments no.7 and 8 are equivalent except the extent to which there is agreement on the hypotheses P(*H*)=1/2 or P(*H*)≠1/2 respectively and P(*B*)=1/2 or P(*B*)≠1/2. One of the major problems confronting a researcher in interpreting summary answers may be sketched within this context. A substantial proportion of students (in our interviews) held the strong hypothesis P(*B*)=1/2 in the bag item; they had falsely assumed equal numbers, trusting that if numbers were unequal, it would be stated clearly. They end up with the result (=) and choose (*B*) or (*W*) as they please. The result of their reflection is the same as in the coin tossing experiment 7. But, in the coin tossing they tend to choose to break the series of heads with an answer of tails, and in the bag item 8 they tend to choose the continuation of the series, namely answer (*B*). Even when directly confronted with this inconsistency in their answers, most of them did not change their mind on that issue.

Teaching. We believe that in pursuing one normative modelling of a problem, teachers often neglect its full complexity. Experiment no.8 is by no means trivial as our discussion of possible strategies reveals. If the complexity increases, learners have to rely on the extrinsic character of the situation and on subsidiary strategies with all the consequences for the effect of teaching already alluded to.

Dealing with probabilistic information is crucial for intuitive understanding of statistical theory. There are hypotheses on coin tossing which many people agree with so that sample information plays a minor role. Other hypotheses are not so strong as in shown in the bag experiment no.8, whence one has to include sample information to perform a test. Still other hypotheses are so vague that no tests of specific hypotheses are sensible and sample information plays a substantial role. Objectivists formalize the available information, be it vague or more precise, very indirectly via the choice of the α-level for a type I error. Subjectivists, on the other hand, evaluate hypotheses directly via probabilities and express the vague information by a uniform distribution on (0,1); they combine this prior information on hypotheses with sample information to new (posterior) information on hypotheses by means of the Bayesian formula.

The comparison of experiments no.7 and 8 illuminates these issues about dealing with varying quality of information. It is important to exploit the differences between related problems as stated in the previous paragraphs. Students may learn best about different but related ideas, about similar model assumptions, and about different but interwoven concepts, by a systematic comparison of related but not identical situations such as items no.7 and 8.

6. Concluding Comments

Empirical research

We restructure those aspects discussed in the experiments into categories which mark crucial points for the evidence of empirical results and at the same time enlighten probabilistic concepts. We concentrated our analysis on the internal ambiguity of the items. External conditions of the situation, however, might heavily influence the answers. Falk et al. (1980, p.196) report how seemingly insignificant details influence the behaviour of subjects; children chose that urn which was nearer to the prize promised if their choice was correct. The external layout of an experiment thus may impede the necessary abstraction process in artificial situations. Besides that, pressure of time may have serious consequences on the problem solving behaviour so that the interpretation of subjects' answers becomes difficult. In our opinion one must allow subjects a long span of time to come to terms with the posed problems to obtain deeper insights.

There are various crucial points leading to difficulties in the communication process between the experimenter and the subject. Seemingly false answers can be traced back to such breakdowns in communication rather than to missing concepts. For instance, subjects may not comprehend the formulation of the problem context if the question is not posed clearly enough. The context may be ambiguous, or the verbalization very similar to the formulation of another problem. Subjects may find it difficult to formulate their problem reconstruction and strategy in a follow up interview. The conclusions drawn by the researchers in such conditions can be rather dubious.

Situations which are classified as artificial or pseudo-real, both have their special difficulty. Artificial context items are mostly intended to be exemplars of abstract theory. They are separated from everyday life experiences even if they are thought to serve as links between intuitions and abstract theory. Thus they necessitate experience of their own. If a learner lacks this experience that does not mean that he/she cannot cope with the related theoretical concept; it only means that he/she is not familiar with that materialization of it and may intuitively make adequate use of the theory in other, more familiar situations.

Pseudo-real context items have to be described by an artificial context or an abstract model before analyzing them. This modelling process may be obstructed by the artificial image not being within the reach of the subject or if the pseudo-real features induce a more relevant reconstruction of the problem. The latter case is of crucial influence for the evaluation of subjects' answers: if the subject is confronted with a pseudo-real item and restructures it according to common sense, then this strategy should not be interpreted as a lack of adequate concepts. It is of interest that even artificial contexts reveal unexpected features.

The situation in many items in empirical research involves a limited probability concept. The vast majority of problems deal with the symmetry concept, few deal with relative frequencies; the inherent dynamics of probability is hardly covered. Despite the emphasis on the symmetry concept there is a conflict with other conceptions of probability as some of the problems are not resolvable with symmetric probabilities only.

There is a gap between concepts and problem solving strategies which reduces the value of the items considerably. For example children might have learned that fractions lead to answers which are evaluated to be correct by the teacher so that they are able to solve some problems without understanding them. Specific answers cannot directly be linked to specific problem solving strategies; it is even more dubious to infer underlying concepts from used strategies. Only feedback on the subject's problem reconstruction and strategy actually used will establish such links.

Ideas are developed to explain and diagnose inadequate behaviour; they bear the danger of acquiring a negative connotation as the representativeness heuristics does. It may be forgotten that they also represent powerful strategies. Furthermore, one cannot expect that patterns of explanation are applicable throughout all relevant cases and that 'misbehaviour' is to be traced back to only some few cognitive strategies. Our scenario of potential strategies should encourage more research in the area.

Our discussion has concentrated on critical aspects of empirical research on the probability concept so that the positive merits of the various authors are not brought out. Each of the projects and papers on understanding probability is helpful. However, our intention has been to initiate awareness of the multifaceted problems involved in such research and to develop more reliable insight into the exciting intuitions on probability.

Teaching consequences

Crucial aspects for interpreting empirical results on probabilistic understanding also mark difficult points for conceptual understanding. Whenever communication in an interview breaks down, such circumstances also decrease sensible communication in class. We summarize the issues discussed for empirical research as a guideline for teaching:

- External conditions of interviews.
- Problem reconstruction and its verbalizing.
- Type of situation involved.
- Spectrum of probability concepts.
- Strategies to solve vs. concepts to perceive a problem.
- Patterns of explanation for misconceptions.

If teaching is aimed at a deeper comprehension of concepts and not only at mechanical strategies then these categories may obstruct the necessary development of the interplay between intuitions and mathematics. Some of the points conflict with common sense, some with verbalization of context and strategies, some with (unconscious) differing intuitions, and some with the kinds of learners' intuitions. All of them deal with possible breakdowns in sensible communication. One comment may be hard to accept: understanding probability is no easier if one tries to teach a reduced concept with a reduced spectrum of intuitions. Didactic efforts mislead if, for the best intention, they are primarily aimed at simplifying matters and thereby miss the complexity of concepts and understanding.

Courses still remain to be developed which emphasize the comprehension of basic concepts and their mutual effects at an intuitive level whilst holding technicalities within the learners' range. Explicit and open discussion of various pupils' solutions in class could establish the interplay between primary intuitions, mathematical concepts, and secondary intuitions; no new examples need to be invented, only the way of dealing with them has to be revised. Intuitions may be more effectively revised by mistakes, diverging problem reconstructions, and different solutions rather than by extrinsic solution of many problems, which are not correct solutions at all as our discussion reveals.

From the didactic perspective it may be advantageous to develop unconventional concepts which allow for a more direct development of stable secondary intuitions on traditional probability concepts. Such approaches are rare. We have tried to elaborate some ideas such as the complementarity between the intuition of representativeness and its formal counterpart of random selection. We have also developed unconventional concepts like 'common sense probability' (Bentz, 1983) and the 'favour concept' (Borovcnik, 1987) which focus on intuitive understanding of formal probability. We have received positive feedback on our own approach of integrating empirical research and unconventional concepts in teaching. A promising approach might be to view teaching in class as empirical research on students' views.

Bibliography

Assessment of Performance Unit (APU): 1983, *Mathematical Development, Primary and Secondary Survey Reports*, 1,2,3, Her Majesty's Stationary Office.

Bentz, H.-J.: 1983, 'Stochastics Teaching Based on Common Sense', in: *Proc. First Int. Conf. on Teaching Statistics*, vol II, Teaching Statistics Trust, Sheffield, 753-765.

Bentz, H.-J., Borovcnik, M.: 1986, 'Zur Repräsentativitätsheuristik - Eine fundamentale statistische Strategie', *Der Mathematisch - Naturwissenschaftliche Unterricht* **39**, 398-403.

Borovcnik, M.: 1985, 'Begünstigen - Eine stochastische Intuition', *Praxis der Mathematik*, **27**, 327-333.

Borovcnik, M.: 1987, 'Revising Probabilities according to New Information - A Fundamental Stochastic Intuition', in: *Proc. Sec. Int. Conf. Teaching Statistics,* Victoria, 298-302.

Cohen, J., Hansel, M.: 1956, *Risk and Gambling*, Philosophical Library Inc., New York.

Falk, R.: 1983, 'Experimental Models for Resolving Probabilistic Ambiguities', in: *Proc. Seventh Int. Conf. Psychology of Mathematics Education*, Rehovot, 319-325.

Falk, R., R. Falk, and I. Levin: 1980, 'A Potential for Learning Probability in Young Children', *Educational Studies in Mathematics* **11**, 181-204.

Falk, R. and C. Konold: 1991, 'The Psychology of Learning Probability', in: F.S. Gordon and S.P. Gordon, *Statistics for the Twenty-first Century,* Mathematics Association of America.

Fischbein, E.: 1975, *The Intuitive Sources of Probabilistic Thinking in Children*, Reidel, Dordrecht - Boston.

Fischbein, E.: 1987, *Intuition in Science and Mathematics. An Educational Approach*, D. Reidel, Dordrecht.

Green, D.R.: 1982, *Probability Concepts in 11-16 Year Old Pupils*, Report Loughborough University of Technology.

Green, D.R.: 1983, 'A Survey of Probability Concepts in 3000 Pupils Aged 11-16 Years', in: *Proc. First Int. Conf. on Teaching Statistics*, vol. II, Teaching Statistics Trust, 766-783.

Hawkins, A.E. and R. Kapadia: 1984, 'Children's Conceptions of Probability - A Psychological Review', *Educational Studies in Mathematics* **15**, 349-377.

Kahneman, D. and A. Tversky: 1972, 'Subjective Probability: A Judgement of Representativeness', *Cognitive Psychology*, 430-454.

Kahneman, D., D. Slovic, and A. Tversky (eds.): 1982, *Judgement under Uncertainty; Heuristics and Biases*, Cambridge University Press, Cambridge.

Kapadia, R.: 1984, 'Children's Intuitions and Conceptions of Probability', manuscript, Polytechnic of the South Bank, London.

Kapadia, R. and J. Ibbott: 1987, 'Children's Subjective Notions of Probability', manuscript, Polytechnic of the South Bank, London.

Konold, C.: 1989, 'Informal Conceptions of Probability', *Cognition and Instruction 6(1)*, 59-68.

Kütting, H.: 1981, *Didaktik der Wahrscheinlichkeitsrechnung*, Herder, Freiburg.

Piaget, J.: 1975, *Die Entwicklung des Erkennens I: Das mathematische Denken*, Klett, Stuttgart.

Piaget, J. and B. Inhelder: 1951, *The Origin of the Idea of Chance in Children*, English translation 1975, Routledge and Kegan Paul, London.

Scholz, R.W.: 1981, *Stochastische Problemaufgaben. Analyse aus didaktischer und psychologischer Sicht*, IDM-Materialien und Studien, Universität Bielefeld, Bielefeld.

Shaughnessy, J.M.: 1981, 'Misconceptions of Probability: From Systematic Errors to Systematic Experiments and Decisions', in: A.P. Shulte and J.R. Smart (eds.), *Teaching Statistics and Probability*, National Council of Teachers of Mathematics, Reston.

Shaughnessy, J.M.: 1983, 'Misconceptions of Probability, Systematic and Otherwise; Teaching Probability and Statistics so as to Overcome Some Misconceptions', in: *Proc. First Int. Conf. on Teach. Stat.*, vol. II, Teaching Statistics Trust, Sheffield, 784-801.

Tversky, A. and D. Kahneman: 1973, 'Availability: A Heuristic for Judging Frequency and Probability', *Cognitive Psychology*, 207-232.

Manfred Borovcnik
Institut für Mathematik
Universität Klagenfurt
Sterneckstraße 15
A-9020 Klagenfurt
Austria

Hans-Joachim Bentz
Fachbereich Mathematik/Informatik
Universität Hildesheim
Samelsonplatz 1
DW-3200 Hildesheim
FR. Germany

Chapter 4

Analysis of the Probability Curriculum

A. Ahlgren and J. Garfield

Recently there has been movement around the world to introduce probability into elementary and secondary curricula, for a variety of reasons: it is required for reading newspapers, being an informed citizen, it helps one to understand and evaluate information in the world around us, and it is prerequisite to entering many other fields of study (Råde, 1985). In the present school systems of the world, students may encounter probability topics, by themselves or in connection with descriptive statistics, as early as age six, as late as upper secondary school, or not at all.

Because of the lack of mathematical sophistication of secondary students and the different views on the nature of probability, there are also very large differences in what is taught. At one extreme there are collections of isolated lessons to be used ad hoc by interested teachers, wherever and whenever they choose. At the other extreme are entire mathematics curricula that include a progression of instruction in probability topics building from Kindergarten to age 18.

Not only have there been few advocates for probability in the secondary school, there are no slots in students' programmes and no one who can teach it. The absence of a tradition for secondary probability instruction, such as is enjoyed by algebra, geometry, or even calculus, makes it possible to ask rather open-endedly what is to be learned and how and when it should be taught. There is the danger, however, of including topics in the curriculum without first determining what we want students to know, why they should know these things, how they will use them, and what is known about the difficulties of designing and implementing new curriculum.

In this chapter we begin by describing the special problem of probability in the curriculum and positing what we want students to end up knowing about probability at the end of their school experience. Next we present some important general issues of what curriculum is and where it comes from, and focus on some concerns about how it relates to students. We then consider alternative forms of the probability curriculum, using current projects as examples of each form. Finally, we summarize the theoretical issues as ques-

R. Kapadia and M. Borovcnik (eds.), Chance Encounters: Probability in Education, 107–134.

tions to be asked about any curriculum, and recommend ways to use current curriculum efforts to extend our knowledge of how students think about and learn probability.

1. General Aims

Objectives

Lists of probability topics for the secondary curriculum have appeared in several contexts, either in the form of recommendations for the curriculum or in the form of actual curriculum materials. The best current example is a set of recommendations prepared by the National Council of Teachers of Mathematics. The following quotations are the core of the recommendations for probability in their report, 'Curriculum and Evaluation Standards for School Mathematics' (NCTM, 1989).

> **Ages 5-10**: Explore concepts of chance.
>
> **Ages 10-14**: Model situations by devising and carrying out experiments or simulations to determine probabilities; or by constructing a sample space to determine the probability of an event; appreciate the power of using a probability model through comparison of experimental results with mathematical expectations; make predictions based on experimental results or mathematical probabilities; develop an appreciation for the pervasive use of probability in the real world. (p. 109)
>
> **Ages 14-18**: Use experimental probability, theoretical probability, or simulation methods as appropriate, to represent and solve problem situations involving uncertainty; understand the concept of a random variable, create, interpret and apply properties of discrete and normal probability distributions, and, for college-intending students only, apply the concept of a random variable to generate and interpret probability distributions including binomial, uniform, normal, and chi square. (p. 171)
>
> Students at this level should understand the difference between experimental and theoretical probability. Concepts of probability, such as independent and dependent events, and their relationship to compound events and conditional probability should be taught intuitively. Formal definitions and properties should be developed only after a firm conceptual base is established so that students do not indiscriminately apply formulas. (p.122)

As will be clear later when research on how people learn probability is considered, the list seems ambitious. At about the same time, instructional materials were developed by the associated Quantitative Literacy project of the NCTM and the American Statistical Association (ASA). These materials only reached as far as the very first part of the Ages 14-18 objectives just quoted.

Another example of the optimism of professors is the Probability unit of the Middle Grades Mathematics Project for ages 10 to 14 (Phillips et al., 1986). The stated goals for the very first activity are that the student will understand the definition of probability, that zero represents impossibility, that 1 represents certainty, that the sum of all probabilities in a situation is 1, and also "understand the definitions of fair and unfair as applied to games or probabilistic situations," "simulate a situation and discuss the results," and "use probability to decide whether a situation (or event) is fair or unfair."

The last decade of research in mathematics and science education has made amply clear that students leave secondary school with very little understanding of mathematical ideas, or ability to apply them to real-world problems, even when they are able to perform algorithms that their teachers had taken as demonstrating understanding (Garfield and Ahlgren, 1988). Just what can be done about the inefficacy of mathematics education may take years to discover, but surely it is evident that we must begin by specifying the requirements for conceptual understanding, rather than merely listing topics or listing calculations that can be made.

Pulling together the ideas concerning the history of probability and how people think about probability from the other chapters of this book, what can we conclude are reasonable goals for what students ought to learn, say, by the time they graduate from secondary school? We will address ourselves in this chapter to what all graduating students ought to know, not just those who are good at mathematics or will specialize in technical subjects. We do not assume that specializing students, by virtue of learning algorithms and more advanced topics, will necessarily have learned these basic ideas. Our focus here is not on what lessons should be taught on some occasion, but on the residual understanding that will remain well after the completion of schooling.

We distinguish between three interdependent domains of outcomes: a coherent set of ideas concerning probability, skills in using these ideas, and the inclination to apply these ideas and skills. Note that we exclude the ability to perform calculations without understanding. The opinion that meaningless operations can be learned and stored for later use, and possibly later understanding, is at best overly optimistic and at worst wasteful and cruel. In a similar vein, Schoenfeld (1982) has found most of the teaching of school mathematics to be "deceptive and fraudulent."

Ideas

By the end of secondary school at the age of 18, students should understand this progression of ideas (see also Rutherford and Ahlgren, 1990; pp.104-5):

1. We like to believe that the world is causal: if we know what the influences acting in some situation are, and if we also know rules for what effects the influences have, then we may be able to figure out just what will happen next. But most events of interest to us are the result of so many complicated causes that we do not know what they all are or cannot measure them satisfactorily. Furthermore, we might not know the rules for what their effects are, or we have not got computers powerful enough to compute the consequences of the rules even if we knew them. For many kinds of events, it may even be impossible to predict in principle, as they are highly sensitive to minuscule differences in the initial conditions (e.g., the butterfly effect on the weather).

2. We can, however, still talk intelligently about how likely we believe some outcome of a situation to be because all the many influences tend to more or less balance one another. We can therefore compute what we might expect for a series of encounters without predicting single outcomes.

3. It is often useful to express strength of belief in an outcome numerically, say on a scale of 0 to 1 or 0% to 100% -- where 0 means we believe there is no possibility of a particular result occurring, 1 or 100% means that we are completely certain that it will, and any number in between indicates there is only that probability of the outcome occurring. If our ideas are consistent, then the sum of the probabilities for all possible mutually exclusive outcomes of a situation must equal 1.

4. In estimating probabilities for the outcomes of a situation, one useful thing to consider is what has happened in similar situations in the past; if the current situation is like some class of situations in the past, we may expect similar proportions of various outcomes in the present.

5. Informal estimates of probability based on experience are often strongly influenced by non-scientific aspects of thought: people most readily take into account what is easiest to remember, what fits their prejudices, what seems special about the current circumstances, and so on. We should beware this tendency in others and in ourselves, for it distorts our judgement of a class of 'similar situations.'

6. Another way to inform our judgement is to think about alternatives for what could happen. If we know what all the important alternatives are, we may better be able to estimate what the chances are for any one of them occurring. If we cannot see why any one of these alternatives is more likely than any other, we might estimate the probability of each event to equal what fraction it is of all the possibilities.

7. Still another source of our strength of belief in a particular outcome is an informal estimate that a preponderance of causes in one direction will swamp unpredictable influences. This sort of analysis, however, is strongly influenced by 'wishful thinking' -- our desire for the one outcome inclines us to ignore contradictory influences and exaggerate consistent ones.

8. When a process is believed to be so unpredictable as to be essentially random, it may be useful to simulate the process using random numbers with random devices or computers, to get an idea of the relative frequencies of different kinds of outcomes.

9. Whenever we are estimating probabilities, we are better off if we ignore any particular single event, but concentrate on what will happen in groups of similar events. Considering only what will happen on the average for the group can be useful in making decisions -- even if it does not satisfy our desire to know about each separate case.

10. Seeing how a situation actually turns out, of course, can affect our beliefs about what influences were acting, or even about what alternative outcomes are possible. But caution must be exercised not to overinterpret a few outcomes. When outcomes are uncertain, it may happen that a well founded decision may fail and a nonsensical strategy may prove successful. When that occurs, people may develop a strong belief that their nonsensical strategy is justified.

11. When there are two or more distinct, alternative outcomes of a situation, the probability of one or another of them occurring is the sum of their separate probabilities. Sometimes when an outcome seems surprising, it is useful to imagine how many other outcomes would also have seemed surprising -- and consider what the sum of all their probabilities would be; e.g., the probability is fairly high every year that at least one small town out of thousands will show a surprisingly high rate of some disease among the dozens on which records are kept. True, for any specific town and disease the probability of such a high rate is extremely small. However, these small probabilities have to be added for all the diseases observed and all the towns which keep records.

12. When there is reason to believe that two or more events have no causal links between them, one may be inclined to estimate the probability of them all occurring by

computing the product of their separate probabilities. But if there is an unsuspected causal connection between the events (one influences the other or both are influenced by some third factor), or just an accidental correlation between them, the estimate can be pretty far off. For instance, the probability that *all* the systems will work properly in launching a rocket into space is usually taken as the product of the individual reliabilities. If however, one and the same influence, say bad quality of raw material, has caused a failure in several of the components, the computed reliability is misleading.

13. On the other hand, when there is a connection between events, it is often useful to specify the probability for event E given the circumstance F. The probability for event E occurring at all is then the product of that conditional probability and the probability of circumstance F. For instance, the probability of being bitten by a poisonous snake is the product of the probability of encountering such a snake and the probability that, if so encountered, it would make a successful bite.

Skills

The ideas behind probability are difficult to learn and therefore hard to teach. One retreat of the instructor is to make a detour around these ideas and to focus on skills such as combinatorial methods which do not contribute anything to the understanding of the ideas. Memorizing difficult skills is a traditional approach in mathematics teaching to avoid discussing the sometimes vague underlying ideas. We believe that a promising alternative is to reduce goals to a reasonable level and leave more time for discussing and developing ideas.

By the end of secondary school, students should be able to:

1. Assign relative and numerical likelihoods to outcomes of events and give some rationale, frequentist, or classical, or subjective, for their judgements.

2. Compute fractions from data series.

3. Judge the extent of 'similarity' in familiar contexts in order to use prior frequencies.

4. Count possible outcomes in very simple combinatorial situations, certainly no more than two dimensions.

5. Make realistic estimates of conditional probability in familiar situations.

6. Consider population size as well as probability in judging risks and benefits; e.g., a probability of only 0.001 that a medical test will yield false results implies 100,000 mistakes in a population of 100,000,000 people.

7. Revise probability judgements qualitatively when given more evidence.

8. Make reasonable judgements about when results should be attributed to special circumstances rather than to base rates.

9. Calculate the fair stakes from probability ratios and vice versa, to see the consequence of probabilistic 'risk' in a more familiar, monetary context.

10. Use tree diagrams as a tool for modelling probabilistic situations.

11. Apply addition and multiplication rules in familiar contexts and discuss when they are not fully appropriate.

Inclination to apply ideas and skills

By the end of secondary school, students should be inclined to actually exercise the skills listed above when occasions arise; for example, when the newspaper claims that three home runs in a row are just short of miraculous, that the odds are three to one that UFO's are real, or that a new drug is 95% safe. Research shows that transfer of training to novel contexts is poor, and such real-world behavior is not likely to occur unless students are given extensive practice in it. Urns and dice are not enough.

2. General Curriculum Issues

Aspects of the curriculum

The term curriculum is used in a variety of senses from, at one extreme, 'a list of course titles' to, at the other extreme, 'whatever ideas and skills students actually happen to learn.' The range of meaning has led to distinguishing among five denotations:

Planned curriculum. This is the written plan for a school system that includes specifications of what is to be taught, in what sequence, and at what ages. There may also be suggestions for methods of teaching and for means of testing students' achievement. The planned curriculum can be assessed merely by reading documents.

Taught curriculum. This is what teachers actually do in classrooms -- what they lecture on, what questions they ask, and what problems they assign. This can be assessed by teacher reports or, better, by classroom observations.

Learned curriculum. This is what ideas and skills the students actually learn from their experiences in class and in doing homework. These can be assessed by testing students at the end of the course, or sections of the course.

Retained curriculum. This is what ideas and skills students still recall some time after instruction is over. For example, how much they remember by the time another course begins in which previous knowledge is assumed, or how much they remember by the time some application actually arises in their extra-curricular lives.

Exercised curriculum. This is what ideas and skills the students actually call on subsequent to the course in which they were taught -- the 'inclination' of the Skills section above. The use could be in their personal lives, their jobs, or -- although there is some danger of circular reasoning here -- in subsequent schooling.

Even among the members of a commission that has planned a curriculum there are likely to be several interpretations of what is intended by the text of the final version of the curriculum. Clearly, the interpretation by different teachers could be different, too. That is because of their backgrounds and the specific classes they imagine in their minds which are more individualistic than the general perspective that curricula planners have to have in their minds. What is taught and how it is taught might influence not only what students learn at the time when they are in class but also what they will retain after some time has elapsed. Furthermore, the individual students, or interacting groups of students learn on an even more individualistic basis and learn somewhat different things even if they are taught in the same class.

There is a subtlety here: the students may only partially grasp ideas at the end of a particular course of instruction, but through a series of planned exposures in different contexts will gradually develop a more complete understanding. In this sense, the eventually retained curriculum may be more than the sum of the immediately learned curriculum. This possibility should not be confused, however, with the optimistic hope that teachers need not teach for understanding at all, for understanding will eventually grow from a progressive accretion of rote learning.

Commonly it is greatly overestimated how much students can apply to their actual lives from what they have learned abstractly in school. Moreover, the inclination to do so decreases sharply with only a formal understanding of what has been learned. To learn less, at a lower level -- but to learn it in meaningful contexts -- might influence students' behavior more directly and effectively with regard to the exercised curriculum than to train high level procedures formally in the taught curriculum.

In this chapter we can appraise only the planned curriculum, and make some suggestions about what to be alert for in considering the locally taught and learned curriculum. The authors share the attitude, however, that schooling efforts that do not lead to useful retention are a waste of everyone's time. Curriculum should be planned with the goal of the retained and exercised curriculum ever in mind.

Traditional curricular thought centres on the planned curriculum, in particular on arguments for the inclusion of topics and distribution of the topics over time. The assumption is commonly made that the retained curriculum will be identical to what is planned. Which means either that the taught-learned-retained-exercised transformations are not recognized at all, or that they are believed to be controllable. Progressive curricular thought tends to focus on the retained and exercised curriculum, that is, on what one might reasonably expect students to remember and on how that would have some benefit in their lives. One usually still ends up with a planned curriculum, but one that takes far more account of how people learn and what their lives are like.

There is some danger of both schools of thought neglecting the uncertainty of the teaching link. In technical subjects, particularly at unspecialized levels of schooling, there is, at least in the USA, a considerable problem of the teacher having little more interest in, aptitude for, or knowledge of the topics that the students do.

Curriculum sources

Propounders of curriculum theory identify a number of different sources of wisdom from which decisions about what should be taught can be derived (e.g., Eisner, 1979). There is nothing necessarily contradictory about the sources, so more than one is likely to be involved even in the thinking of any one educator when actual curricula are developed. Debate about particulars can be very confused, however, unless the distinct sources are acknowledged.

Academic disciplines. It is characteristic of the traditional approach to curriculum that the source of wisdom lies in the organized knowledge of the academic disciplines. Thus, for example, the rationale for a stochastics curriculum would be seen as lying in mathematics. The knowledge to be learned, the skills to be practiced, and the logic of instruction are mathematical. The distribution of topics over time will be governed by the development of students' computation skills. The knowledge and skills to be learned will be of greatest subsequent use on mathematics examinations, although it is commonly believed that they will thereafter lie ready to be used in case any practical application should arise.

The logical consistency of the subject matter will be the focus of most instruction. Because the abstract mathematical way of defining concepts is difficult to grasp, coins, urns, and the like are used in an attempt to make the ideas concrete. Real world applications typically are viewed not as providing a rationale for study, but as 'examples' of mathematics. Unfortunately, it is difficult and unappealing for students to apply ideas learned in the abstract to real problems later on. Many research findings show that the transfer into new contexts is remarkably poor unless extensive training in doing just that is provided. An academic discipline orientation implies that probability should get a bigger weight than statistics which should then be based on probability concepts rather than on data analysis. Greater importance would be given to mathematical theory, to precise language, and to the description of events in terms of sets.

Nature of the world. A position similar to a discipline-oriented one purports to draw on the very existence of some aspects of the world as a source of curriculum. Essentially this consists of choosing some part of existence that is particularly interesting to the proposer (time, dust, markets, uncertainty, etc.) and citing the pervasive existence of this aspect to justify the necessity of all students studying some discipline that deals with it. Happily, there will always be such a discipline available to satisfy the task of describing the nature of things, because it is only through the disciplines that we are likely to have identified teachable aspects of the nature of the world in the first place. Thus the nature of the world is only a pretence to introduce a covertly disciplinary approach. The organization of instruction would be similar and the problems with transfer of acquired knowledge to real problems would be the same as in the disciplinary approach, because the nature of the world was not explicitly the topic of instructional efforts.

Needs of society. Another important source of wisdom is the observation of what the social system needs by way of abilities in its citizens. If voters and workers can be seen to require some set of skills and understanding to be a valuable part of society, to vote wisely and work well, then surely the responsibility of the school system is to produce those. Thus, for example, the rationale for a stochastics curriculum would be seen as lying in the need to understand reports of surveys and studies and to make intelligent decisions in uncertain situations. The knowledge would be mathematical, but the skills would be making sense of reports and making sound decisions.

Consequently, teaching focuses on structuring real situations. One of the problems with this is to how to balance between the required knowledge from the context and knowledge from mathematics. If one learns at best only what one practises, students would have to read reports and make decisions, drawing on mathematical references only when appropriate. This orientation lays less weight on the structure of the discipline which is only developed to that extent as is necessary to understand the actual decision. Probability thereby loses its leading role as the only basis of statistics; the general logic of argument becomes more important than its exact probabilistic justification. The distribution of topics over time will be constrained somewhat by the students' development of computational abilities, but will be determined mainly by their sophistication in understanding the context and the logic of the argument involved.

Psychology. Another, more recent view of the source of curriculum is the study of how people think and learn -- i.e., cognitive and developmental psychology. It requires some knowledge of how students typically think about things, how their thinking changes, and how the rate or direction of change can be affected. The curriculum is then designed to move the students from where they are, along anticipated paths, toward where we would like them to be. Thus the rationale for a curriculum is seen as lying in the naive ideas students have about uncertainty and the feasible kinds of changes in their thinking that could be induced.

For example, a developmental psychologist, believing that most students' spontaneous ideas of probability are not frequentist but subjectivist, might conclude that the curriculum should focus on enhancing the accuracy of students' subjective judgements. If the developmental psychology of probability happened to be correct (and there is no guarantee that it will be), then the students would actually learn what was taught in the curriculum. Teaching would focus on organizing students' thinking processes with the help of mathematical concepts. Probability would have more central interest in this approach, because it would be a basis for making more successful judgements. Goals for curricula are more modest than those from a mathematics perspective and they are more consistently reachable.

Teacher knowledge. In addition to disciplines, social benefits, and student psyches as sources of wisdom in curriculum, there is also the possibility of considering the nature of teachers and teaching. That is, one would find out how teachers themselves think -- which, after all, is what they will try to teach the students -- and build the curriculum from there. Although there is current research being done on teachers' ideas, the implications envisioned seem to concern the training of teachers; and, as students, the teachers will have the same problems everyone else does.

In the major curriculum reforms of the 1960s, attempts were made to develop 'teacher-proof' materials that would teach students well regardless of the teacher's capabilities. The strategy did not work because the teachers' own knowledge, perceptions, and attitudes have a powerful filtering effect on what reaches their students. Part of the filter is surely cognitive, but part is also censorial -- at least in the USA where teachers have a great deal of control over what actually happens behind their classroom doors, regardless of what the 'official' curriculum is in the school. In other countries, where there is great central control of curriculum, acceptance by teachers may not be so serious a problem (e.g., Nemetz, 1983).

Tradition. As a final note on sources of curriculum, we should acknowledge the single most powerful influence on what is taught in schools today: what was taught yesterday. Whatever the rationale for yesterday's curriculum may have been, that curriculum's very existence shapes our ideas of what schooling is supposed to be like. It also defines the knowledge that scholars and authorities base their distinction and prestige upon, and that many feel they must defend. This traditionalism is not the same as derivation from the academic disciplines, which themselves change rapidly.

There obviously is a lack of a tradition of teaching probability. In calculus, by contrast, there have been several cycles of teaching with feedback, so substantial experience has been acquired with the design of suitable examples and so on. Probability teaching needs time to acquire the same amount of experience. The major vehicle of the past is the textbook. For a previously untaught subject, however, publishers suffer great uncertainty, and may have to look to nascent guidelines from various mathematical 'authorities'. Their recommendations for distribution of topics over the grades may arise from logical considerations alone, unconstrained by the mathematical limitations of children. For example, an early draft of curriculum guidelines from the American Statistical Association (ASA, 1985) recommended for 13-year-olds topics that are difficult for upper secondary students.

Choice of orientation

It is clear that the starting point for including probability in most current curricula is the nature of mathematics, not that of students' minds or their world. Witness that the teaching of probability has occurred almost exclusively in regular mathematics classes, not in social studies, health, or natural sciences classes where it would have considerable relevance and where it might be learned more meaningfully. One has not asked, 'What do people need to know about stochastics?', or 'How do people actually think with regard to stochastic phenomena?' or 'How can the ideas involved in probability be more logically connected to one another?' but rather 'How can we get people to learn principles of mathematical stochastics?'

Certainly, the nature of students and their world is seen as relevant -- because the former constrains instruction and the latter provides examples. But note that the orientation -- real-world problems are not seen as the source of the instruction, but as instructional devices to help teach principles for their own sake. There is an essential distinction to be drawn between 'applications' problems, which come at the end of each chapter so that students know what mathematics is to be used, and 'modelling' problems, which come at the beginning so that the mathematics must be drawn from them. The extent to which the end-of-chapter problems work as an instructional strategy is a matter for research, not philosophy. The point here is the motivation of the curriculum designer.

Consistent with this view, a detailed specification of curriculum would not be based on the kind of problems students are likely to be facing as they age, nor on their psychological development, but rather on the logical development of the mathematics. Steinbring makes the case in Chapter 5, however, that the logic of the mathematics alone is inadequate to deal with the didactics of probability as probability covers quite a variety of meanings, some still disputed in the literature. Theoretical probability includes deductions from premises that may be inappropriate to the context, or about whose appropriateness we are ignorant. Empirical probability draws on previous conditions that may no longer prevail, or about whose relevance we are ignorant. Neither a theoretical nor an empirical approach to probability is tenable for teaching by itself. So some parallel development is required in which each view illumines the other.

We suggest that the situation is more indeterminate still, because the theoretical and empirical views are not adequate even together. The academic field itself is split several ways on the nature of probabilities: can a true probability for a future event be said to exist in some objective sense, or is it only a matter of our rationalized strength of belief? And, whatever the philosophical ontology of probability, educators must face the largely subjective view of probability held by students -- which includes elements of interest, mathematical skill, misconceptions, and utility.

3. Curriculum Issues in Probability

Issues that are particular to the curriculum on probability concern both the nature of the students and the form that the curriculum takes.

Student readiness

A successful curriculum in probability will have to address four major issues: students must be interested in probability, they have to have the mathematical background to make sense of it, their naive ideas have to be engaged rather than overridden, and transfer to their activities outside of school has to be fostered.

Be sure students are interested in probability. In Konold's view (1989), based on extensive interviews of college students, subjects do not attend at all to series of events, but only to real outcomes of single events. For example, students were given this account of weather-forecasting:

> A meteorologist states on 10 different days that there is a 70% chance of rain that day. On 7 of the 10 days it actually rains; on the other three it does not. How well is he doing?

Many of them answered that the meteorologist was right only on 7 out of 10 days. The specification of 70% was taken as a prediction on each day that it would rain -- so the forecast was right on 7 days and wrong on 3. The students with this "outcome orientation" believed that the forecaster would have been vindicated better if there had been rain on all 10 days! Konold found that students regarded anything much over 50% as indicating something *would* happen, anything much under 50% as indicating it would *not* happen, and 50% as indicating complete ignorance. Lest it be thought that their judgements were unusual because of the obvious personal consequences of preparing for rain, Konold found similar results for urns with a 5 to 1 ratio of black to red balls; they judged 'all black' as the most likely sample of 6 marbles -- because on every draw, black is much more likely than red. The fundamental misconception is shown dramatically by a student's comment (Konold, 1988):

> "I don't believe in probability. Just because the probability of something is 40% doesn't mean that it can't happen. Even if the probability is only 2%, it can still happen. So I don't believe in probability."

Moreover, Konold (1989) found college students not to relate numerical specification of probability directly to observed frequencies or to theoretical counts of possibilities. They would use empirical or theoretical information or causal propositions to form a strength of belief first and, if they were required to make a numerical statement of probability, they would subsequently derive a number from their strength of belief and not from the quantitative information given.

The claim that students think about probability subjectively does not imply that the students are subjectivists in a formal sense. Indeed, the subtleties of Bayes' formula with the systematic revision of prior probabilities on the basis of new information are well above the heads of most of them. The recent recognition of this subjective conception of probability now makes it possible to pursue research on how to move the student towards the formal ideas of probability, but we do not know much about that yet.

As it stands at the present, most curriculum in stochastics is a pot pourri of theoretical, frequentist, and subjectivist ideas, not carefully articulated but presented haphazardly as if they were all the same thing. The students' interest in this is likely to be minimal. Moreover, each student's emerging idea of probability is not entirely consistent with any one of the technical meanings. Many people tackle problems with an idiosyncratic reconstruction of a problem; and they have often good reasons for doing so. Instruction, however, usually restricts problems artificially and suppresses most intuitions. Everything which has no clear place in the formal theory is ruled out. In this viewpoint, readiness would be thought of as merely the ability to absorb new information. Instead, teaching has to focus on how theory is different from subjective intuitions and why it is still useful to learn these theoretical ideas.

Be sure students have the mathematical background to understand probability. Many students probably do not. In particular, there is a considerable amount of current research on how students learn -- or do not learn -- proportionality. People tend to linear, additive thinking, and there are as yet no breakthroughs on how to develop their proportional thinking (Tournaire and Pulos, 1985). The concepts of combinatorics and relative frequency both assume a grasp of proportionality. Which means that even if we could make students inclined to take classical or frequentist views of events, they might not be able to handle the concepts quantitatively. The teaching of probability may have to await progress in teaching proportionality.

But there are other very basic mathematical ideas that students need to understand probability. For example, the idea that numbers can be used to represent abstract variables -- such as strength of conviction. The scale used for probability is particularly ill suited for naive students, compressing as it does the entire universe of opinion into a tiny range between the first two counting numbers. Indeed, the 0 to 1 convention reduces most of the strong probability statements (very high or low odds) to vanishingly small decimal differences near 0 or 1. The most popular form for expressing strength of belief may be on a scale of one to ten. Percentages are also reasonably familiar. Yet even seeing the equivalence of these different forms requires a fairly sophisticated mathematical perspective. Common use of expressions such as "He gave 150% effort" or "I support that idea 1000%" indicate that 100% is not connected so much to the idea of a fraction as to the idea of a rating scale -- on which one can raise the ceiling for exceptional experiences.

A common curricular answer to concern about the difficulty of an idea is to teach the rote use of it first and hope that understanding of what it means will develop eventually. A familiar example is the prolonged attempt to teach rote performance of the division algorithm. Except for the few students who take to it, the algorithm never becomes a reliable tool and the effort is a source of distress for years of schooling.

An alternative approach was taken in the curriculum developments of the 1960s. In that movement, mathematicians built curriculum on set-theoretic deduction of arithmetical operations, so that children would understand, rather than merely memorize, the algorithms. The movement failed: besides the children's difficulties, neither their teachers nor their parents understood what was going on. Of course there are some things about division that children know and can learn early, but deduction from axioms they did not understand was no more helpful than rote learning.

Furthermore, our understanding of probability, like that of the closely related idea of randomness, is hypothetical, or circular, or both. If proportionality is difficult for many students, consider how much more difficult will be the mathematical sophistication needed to accept that probability is only a heuristic, which like any mathematics may or may not be useful in dealing with any particular phenomenon.

Be sure students' naive ideas are engaged. One inference from the finding that students have subjective notions of probability may be that the initial pedagogical goal for such students should not be to replace or overlay their natural propensities (which will likely only go temporarily underground). Rather, the goal should be to produce changes in these very ideas, helping students to derive their strengths of belief with some consistency from theory or observation. The precise skills of how to modify one's beliefs quantitatively on the basis of new evidence can be developed later. One proposition by Dinges (1981) seems to be consistent with this background:

> "I should like to propose to begin stochastics instruction always with decisions under uncertainty naively conceived by the learners. Subsequently, they are to formulate the procedure of decision. Finally, there is a check how far mathematical models can be developed to criticize the procedure of decision."

Compare this focus on students' intuition about events with the semantic focus of the Middle Grades Mathematics Project (Phillips et al., 1986), in which the teaching script begins, as the first words introducing the first activity, "Have you ever heard of probability? What does it mean?"

Freudenthal's (1973) reaction to the difficulty of formal ideas of probability has been a plea to leave them out of the school curriculum altogether:

> "I like probability and I teach it to college freshmen, but I must say that I would hate it to be introduced into high school teaching because I fear it too will be spoiled, as it has been in quite a few experiments that have been carried out. I know that many probabilists and mathematicians share my view." (p.592)

But Freudenthal recognizes alternatives to a formal approach. He especially endorses Engel's ideas of beginning very early in school with a variety of experiences in stochastic processes, claiming that

> "in this abundance of concrete problems probability is experienced as a part of life. Multifarious and abundant experience in probability -- without mathematization -- may inculcate a modified intuition about how chance phenomena occur, and so facilitate a mathematical treatment in later grades, perhaps preferably in college." (p.613).

Be sure students apply probability outside of school. Another impediment to meaningful curriculum is the tendency on the part of students to compartmentalize what they learn in school. With no practice in retrieving ideas or employing skills outside of the classroom context, that is, with no knowledge of how and when to use them, students are unlikely ever to think again of most of what they learn in school. This is even more true for ideas that are inconsistent with their intuitive ways of viewing the world.

Freudenthal's idea of continual practical exposure to stochastic phenomena might serve in part to counter compartmentalization. Much greater benefit could come, however, from rooting instruction in real-life issues rather than in the artifice of arbitrary games. Coins, dice, urns, and spinners may have use as models for random processes but they offer inadequate motivation for requiring people to learn probability. Memorized techniques for combinatorics or conditionalities are less important than being able to deal with questions of relevance: How predictive are events recorded in the past for events in the future? Are events really independent? Are expectations for a population really relevant for the singular cases in which most people are interested?

The struggle to base probability instruction on meaningful problem situations is evident in the 'Standards' for school mathematics curriculum quoted earlier (NCTM, 1989). Witness first the exhortation that follows the standards statement for ages 10-14:

> "The study of probability in grades 5-8 should not focus on developing formulas or computing the likelihood of events pictured in texts. Students should actively explore situations by experimenting and simulating probability models... Probability is rich with interesting problems that can fascinate students and provide settings for developing or applying such concepts as ratios, fractions, percents, and decimals." (p. 109)

And then witness what form 'interesting' problems take a page later:

> "At the K-4 level [ages 5 to 9], students may flip coins, use spinners, or roll number cubes to begin their study of probability. At the middle-school level [ages 10 to 14], such experiments can determine probabilities inherent in more complex situations using simple methods. For example, if you are making a batch of 6 cookies from a mix into which you randomly drop 10 chocolate chips, what is the probability that you will get a cookie with at least 3 chips? Students can simulate which cookies get chips by rolling a die 10 times." (pp. 110-111)

Yet the final suggestion does offer some hope:

> "With a computer or set of random numbers, this same problem can be extended to simulate an industrial quality-control situation to analyze the number of defective items that might occur under certain conditions on an assembly-line production." (p.111)

Even this, however, is far from what will fascinate students and make mathematics come alive, much less affect their thinking outside of school. Unless probability concepts are taught in contexts of disease transmission, medical testing, accident rates, criminal prosecution, state lottery payoffs, extra-sensory perception, weather prediction, visitors from space, and other 'real-world' matters of real interest, putting probability in the curriculum may not be worth the effort.

Different approaches to probability curriculum

The school mathematics curriculum is overcrowded. One reason is that topics rarely are dropped from the curriculum. An illustration of this is that a high school teacher recently implored a curriculum study group, 'Please, if you do nothing else, make a strong statement about what we can stop teaching in mathematics.' Another reason for crowding is the tendency to press topics into ever lower grades, so that more serious mathematics can be addressed in the upper grades. And the rapid recent developments in mathematics itself motivate the introduction of additional, modern topics such as computer programming and discrete mathematics.

Probability is one of the proposed additions to the crowded curriculum. A case can be made that probabilistic problems provide opportunity for learning other concepts like fractions or proportionality, or for practising computational skills in ways that could be particularly meaningful -- and could therefore be introduced without taking any more time in the long run. But until we know better what it takes to teach probability well, the case is very weak.

Proponents of probability in the schools may have to propose that other topics in mathematics can be neglected. Apparently this has happened successfully in Germany, where students in grades 11 and 12 can choose topics in probability and statistics at the cost of some parts of calculus, linear algebra, and analytical geometry. Another possibility, of course, is that mathematics as a whole should have a larger share of time in the curriculum. This view seems to be prevailing in many who are proposing probability curriculum, at least insofar as they do not specify what other topics in mathematics should be replaced, or what specific economies might be realized in other topics.

On the basis of size, coherence, and context of the curriculum, we identify five basic types of approach:

Ad hoc activities. This consists of utilizing the many suggestions, mostly in the mathematics education literature, for incorporating various games and other activities relating to probability into the regular mathematics classroom. Topics are not formally connected to one another or to topics in statistics. Motivated teachers choose from the materials and use them intermittently in the mathematics classroom. Suggestions for activities appear in the National Council of Teachers of Mathematics (NCTM) Yearbook, *Teaching Statistics and Probability*; in the *Statistics Teacher Network*, a newsletter edited by the joint committee of the NCTM and American Statistical Association; and in the journal *Teaching Statistics* or its German equivalent *Stochastik in der Schule*. It seems unlikely, however, that ad hoc activities would have the coordination of theoretical and empirical viewpoints that Steinbring believes important to optimizing the understanding of probability.

Supplementary units. This approach is characterized by one or more complete but isolated units, typically for two or more weeks' work, intended to supplement regular curriculum, usually in mathematics courses. The units are more comprehensive than a collection of activities, and are likely to have a definite, cumulative sequence. Often topics in statistics are included as well. An example of this approach is the Quantitative Literacy Project. Mathematics teachers receive training in how to teach the materials, which are geared to secondary level students. The project has produced a set of four books for students of roughly 12 to 18 years of age, with no firm designation of the grade in which they might best be used. The books must be used sequentially.

> *Exploring Data* (Landwehr and Watkins, 1986) does just that for a wide variety of real-world data sets, without reference to probability.
>
> *Exploring Probability* (Newman et al., 1987) begins each topic with games of chance, followed by excursions into real-world examples. The motivating topicality of the data sets is largely lost here in the interest of dealing with well controlled circumstances. The persistent reliance on known probabilities camouflages real applications and perhaps decreases motivation.
>
> *The Art and Techniques of Simulation* (Gnanadesikan et al., 1987) is concerned mostly with calculating (by hand or by computer) outcomes of a series of dichotomous events for which, again, the probability is always given.
>
> *Exploring Surveys and Information From Samples* (Landwehr et al., 1987) provides students with elaborate tables of box-plotted confidence intervals for proportions so they can interpret a variety of two-way dichotomous relationships.

Another example of supplementary units is the *Statistics in Your World* series (1980) prepared by the Schools Council Project on Statistical Education in the UK. It comprises 27 pupil books, each "covering a specific topic with the emphasis being on data and on experiments with the pupils taking an active part", and a corresponding set of teachers' notes and pupils' resource sheets. The intended curricular flexibility of the strategy is clear in the publisher's statement:

> "Each book stands on its own -- schools do not have to buy the whole series as an integral course. The books cater for average and above average ability in the first four years of secondary schools, although there is no reason why some of the topics cannot be introduced at an earlier age

given adequate background knowledge; some books have also been successfully used with older pupils."

There are some distinct shortcomings in the use of any ad hoc curriculum or supplementary units. One is that mathematics teachers often have very little familiarity with probability, and so are both unskilled in teaching it and ill inclined to do so. Both projects have been at pains to provide training to the teachers using the trial materials. A second shortcoming is the difficulty of determining the readiness of students for any particular topic. Although this is a difficulty with any curriculum, the lack of continuity in using supplementary units makes the difficulty still worse. It might be worthwhile to attach readiness tests to the units -- or at least caveats to the teacher about what skills students ought to have before approaching them. A third shortcoming arises from the tendency to compensate for the lack of continuity by using materials of high current topical interest as they may lose their interest fairly rapidly and perhaps need to be periodically updated.

There has been little market impact of the *Statistics in Your World* project. The Quantitative Literacy project is reported by the ASA to be used in 3000 schools in the US, but an evaluation study indicates that 'use' may mean only one of the four units, or only selected activities (Garfield and Ahlgren, 1987).

Integration with mathematics. Some complete mathematics curriculum projects have produced materials that include probability topics for all precollege levels. *Real Mathematics* (Willoughby et al., 1987) and the *University of Chicago School Mathematics Project* (UCSMP, 1987) are two recent examples. Here probability topics are included and built upon from elementary through secondary school, in regular mathematics text books and are supposedly coherent with the development of related ideas like fractions in the rest of the mathematics curriculum.

Real Mathematics includes:

Kindergarten: 'Keeping count of objects or events'
'Exposure to probabilistic ideas through the use of games, activities, and stories'
Grade 1: 'Recording frequencies with tally marks'
Grade 3: 'Predicting the outcome of repeated random events'
Grade 4: 'Fractions and probability'
'Coin and cube probability problems'
Grade 5: 'Percents and probability'
Grade 6: 'Odds'

Grade 7: 'Polyhedra and probability'

Grade 8: 'Permutations and combinations'

In the Chicago project, probability is taught informally in elementary grades (ages 6-12); the unit 'Fractions and Probability' is part of the first algebra course for 14-year-olds (McConnell, et al., 1987); the unit 'Probability' is part of the course *Functions and Statistics with Computers* for college-bound 17-year-olds (Rubenstein, 1987).

In the UK's School Mathematics Project, probability is dealt with in a fairly standard mathematical way with emphasis on fractions and equally likely outcomes and there are very few references to chance in realistic situations (Kapadia, 1980). However, they are now producing some new materials on data-handling.

It is common for mathematics textbooks in the USA to include sporadic sections on probability in what can best be called a token fashion. Their appearance allows the publishers to claim to be up to date, but the treatment is typically too abbreviated and unintegrated to be considered anything other than ad hoc. Continental European text-books typically include chapters on probability, but teachers' lack of experience in the subject limits how much is actually taught.

Integration with other disciplines. This approach introduces probability in conjunction with statistics in areas other than mathematics. The Schools Council Project was designed to use this approach.

> "The project's materials link together the statistics used and taught in different subject areas -- social science, geography, humanities, economics, science and mathematics, and help pupils to develop an insight into the statistical problems of everyday life."

However, it has had very limited success in getting non-mathematics teachers to use the materials. In the USA, the importance of teaching mathematical ideas in the context of other subjects is being emphasized in the curricular recommendations of the Mathematical Sciences Education Board (MSEB, 1989) and the American Association for the Advancement of Science (Rutherford and Ahlgren, 1990).

Separate course. In some countries such as Japan, probability combined with statistics is included in a single course in secondary school. The major contents of the course are arrangement of data, basic laws of probability, probability distributions, and the idea of statistical inference. The course is optional and is taken by about half of senior high school students. In the USA, scattered schools attempt about the same as the Japanese.

These courses seem to be gentler versions of typical college courses -- to wit, the premature mathematization that Freudenthal bewails.

4. Approaches to the Probability Curriculum

The road to probability in the curriculum is hardly clear. No matter how well designed a particular curriculum is, its success depends ultimately on the receptiveness and adaptability of the classroom teacher. This lesson was painfully learned by mathematics curriculum developers in the past (Howson, Keitel and Kilpatrick, 1981). We also need to acknowledge that we do not yet know the best way to teach students probability and therefore need to consider several issues before choosing curricula or advocating a particular approach. Moreover, probability is in competition with many other new mathematical topics to be included at school levels earlier than previously thought possible.

Various approaches to probability curriculum are being used around the world. But there is not yet research to show that one method is better than another (or even any good at all) in helping students to solve probability problems or to apply probability ideas to the real world. The chief finding from research is that there are many distortions in how people make judgements involving chance. What is still needed is a common framework by which to evaluate approaches.

This section therefore does not propose a specific role for probability in the curriculum. Instead, we try to capture the many issues discussed in the form of a set of questions that should be asked in considering the construction or adoption of any curriculum proposal. These issues focus on students, topics, theoretical orientation, implementation strategy, instructional methods, and assessment of student learning. Evaluation of the effectiveness of a curriculum should include these issues.

What to look for?

Student readiness. How much does understanding each topic depend on having understood the previous topic? To what extent does each block of lessons provide review to students who have not been exposed to the earlier blocks? Are the mathematical prerequisites clear and what provision is there for assessing them (description of mathematics skills? pretest?).

In projects that integrate probability into mathematics, these questions would be answered by starting students at the beginning of the materials and continuing them through the duration of their precollege schooling. But what happens to students who, as a result of new textbook adoption by the school or transfer to a different school, begin using the materials in the middle years? This is also an issue that applies to separate-units curricula, unless the units are conceived as standing totally by themselves.

Appropriateness of material. How appropriate is the material to children at any given age? Not just can it be learned at the moment, but why should it be learned? How will it be used outside school and how will it be later used in school? How does the new material affect students' thinking about other things in their world? Are real problems used to motivate student interest in the questions, to assist in transfer, and to illustrate useful applications? Are the questions asked likely to evoke students' interest in the answers? In particular, is interest in series of events motivated?

Often, purported 'applications' are just window-dressing for abstract, well defined exercises. We wonder about what motivates the writer of muddled questions like this: "In sets of twins, are 2 boys, 2 girls, and 1 of each equiprobable?" (Answer assumes four equiprobable cases.) Forcing a 'motivating' human-gender dressing on this coin-flipping problem, when the actual probabilities are far different, does more harm than good in fostering interest in probability.

Views of probability. How are frequentist, theoretical, and subjectivist views related? Are assumptions necessary for the validity of each application considered explicitly? In the Schools Council series of supplementary books, for example, probability is introduced empirically (via games of chance), then linked to theory (still games of chance), and then used to model less controlled phenomena. The publisher's description is lucid: in the first relevant book, pupils carry out experiments to find out whether it is "harder to get a six than any other number"; the next relevant book "introduces basic ideas of [theoretical] probability in order to shortcut detailed calculations" in deciding whether a coin is fair. But how do you know when the shortcut is valid? In the next book, the central probability theme is "using simulation to model real-life situations" like booking coach seats, or predicting the weather.

The Quantitative Literacy project also begins with empirical relative frequencies and then introduces theoretical equiprobability. The position is then taken that if you know events are equiprobable, compute the theoretical value of the probability; otherwise, use relative frequency to estimate it. But how do you know? The appropriateness of extending assumptions, such as independence, to the real world is considered in many, although not all, of these excursions.

The Middle Grades Mathematics Project uses the opposite strategy. Probability is defined as a fraction -- "the number of ways A can occur/the total number of possible results." Later we have an experimental probability: "If we get the totals for the numerator and the denominator of our probabilities by counting frequencies from our experimental results, we refer to these as experimental probabilities." In their eagerness to use probability partly as a vehicle for other mathematics topics, principally fractions and their sums and products, this project appears to demonstrate how potentially confusing the logic of presentation can be.

Teacher readiness. How are teachers expected to learn the content of the student materials? What continuing support is offered? History has shown that curriculum innovation is most likely to be sustained if teachers are well indoctrinated and given continuing support and encouragement by the innovators. Training sessions of a few days or even a couple of weeks are usually found to be inadequate (Garfield and Ahlgren, 1987). The Quantitative Literacy project and the Chicago project have set up extensive teacher training programs and users' conferences to deal with this issue. A mathematics curriculum group of the 1960s insisted that every adopting school have a real mathematician attached (Wooton, 1965). Should the same be done with statisticians?

Evaluation. Does the curriculum provide any assessment criteria, guidelines, or materials to help the teacher judge whether the students are ready for new topics or are learning successfully? Does assessment include whether students are understanding the questions and not just producing the correct answers?

Research needs

As is evident in the discussion of many issues raised in this chapter, it is clear that we do not yet know many answers about how best to teach probability to students at the secondary level. There is very little persistent research on it. While many investigators have undertaken studies of difficulties, and on that basis conceived means to improve learning, they have a tendency thereafter to elaborate and promote the means, without further research.

While bright ideas for improvements in instruction should surely be tried, it may be too early for research to provide a sound and extensive basis for improvement. First there must be cross-sectional studies of how students think -- not just whether they get the right answers -- and from those studies estimates made of how the understanding of probability develops. These estimates should be corroborated and refined by longitudinal studies in which actual changes in individuals' understanding are documented. As a reliable research base accumulates for how students think and change in their thinking, instructional methods should be tried through a sequence of alternating revisions and investigations.

What is needed are extensive, cross-disciplinary research projects of how students' thinking about probability develops and how that development can be modified, studies that draw on all of the different sources of probability curriculum (statistics, mathematics, psychology, etc.) This research would focus on the effectiveness of the different curriculum and instructional strategies and materials at all precollege levels. The best chance for this may be within development projects, where commitment to introducing effective practice is initially strong. (After a while, the commitment tends to fix on the current product.) Efforts to develop curriculum and instruction in probability should involve unhurried exploration of the efficacy of their methods in affecting students' thinking.

This is not to imply, however, that research alone can determine efficient educational practice. Research provides information about connections and constraints which the curriculum maker can ponder, and provides feedback about the efficacy of the design. But the design itself requires mathematical and pedagogic creativity that draws on other sources of art and wisdom.

Bibliography

American Statistical Association: 1985, *Guidelines for the Teaching of Statistics within the K-12 Mathematics Curriculum*, Joint Committee on the Curriculum in Statistics and Probability of the American Statistical Association and the National Council of Teachers of Mathematics, Washington DC.

Dinges, H.: 1981, 'Zum Wahrscheinlichkeitsbegriff für die Schule' in W. Dörfler and R. Fischer (eds.), *Stochastik im Schulunterricht,* Hölder-Pichler-Tempsky, Vienna, 49-61.

Eisner, E.: 1979, 'Five Basic Orientations to the Curriculum', in *The Educational Imagination,* MacMillan Co., New York.

Freudenthal, H.: 1973, *Mathematics as an Educational Task,* Reidel, Dordrecht.

Garfield, J. and A. Ahlgren: 1987, 'Evaluation of the Quantitative Literacy Project: Teacher and Student Surveys', American Statistical Association, Washington DC.

Garfield, J. and A. Ahlgren: 1988, 'Difficulties in Learning Basic Concepts in Probability and Statistics: Implications for Research', *Journal for Research in Mathematics Education* **19**(1), 44-63.

Gnanadesikan, M., R. Scheaffer, and J. Swift: 1987, *The Art and Techniques of Simulation,* Dale Seymour, Palo Alto, California.

Howson,G., C. Keitel, and J. Kilpatrick: 1981, *Curriculum Development in Mathematics,* Cambridge University Press, Cambridge.

Kapadia, R.: 1980, 'Developments in Statistical Education', *Educational Studies in Mathematics* **11**, 443-461, .

Konold, C.: 1988, 'Communicating about Probability', Presentation at Research Presession of the annual meeting, *National Council of Teachers of Mathematics*, Chicago.

Konold, C.: 1989, 'Informal Conceptions of Probability', *Journal of Cognition and Instruction,* **6**, 59-98, .

Landwehr, J. and A. Watkins: 1986, *Exploring Data,* Dale Seymour, Palo Alto, California.

Landwehr, J., J. Swift, and A. Watkins: 1987, *Exploring Surveys and Information from Samples,* Dale Seymour, Palo Alto, California.

Mathematical Sciences Education Board: 1989, *Everybody Counts: A Report to the Nation on the Future of Mathematics Education,* National Academy Press, Washington DC.

McConnell, J., S. Brown, S. Eddins, M. Hackworth, L. Sachs, and E. Woodword: 1987, *Algebra,* University of Chicago School Mathematics Project, Chicago.

National Council of Teachers of Mathematics: 1989, *Curriculum and Evaluation Standards for School Mathematics,* The National Council of Teachers of Mathematics, Inc., Reston, VA.

Nemetz, T.: 1983, 'Pre-University Stochastics Teaching in Hungary', in V. Barnett (ed.), *Teaching Statistics in Schools throughout the World*, International Statistical Institute, Voorburg, 85-112.

Newman, C., T. Obremski, and R. Scheaffer: 1987, *Exploring Probability,* Dale Seymour, Palo Alto, California.

Phillips, E., G. Lappan, M.J. Winter, and W. Fitzgerald: 1986, *Probability,* Middle Grades Mathematics Project, Addison Wesley, New York.

Råde, L.: 1985, 'Statistics', in Robert Morris (ed.), *Studies in Mathematics Education,* UNESCO, Paris.

Rubenstein, R., J. Schultz, and J. Flanders: 1987, *Functions and Statistics with Computers,* University of Chicago School Mathematics Project, Chicago.

Rutherford, F. J. and A. Ahlgren: 1990, *Science For All Americans*, Oxford University Press, New York.

Schoenfeld, A.: 1982, 'Some Thoughts on Problem-Solving Research and Mathematical Education', in F. Lester and J. Garofalo (eds.), *Mathematical Problem Solving: Issues in Research*, Franklin Institute Press, Philadelphia.

Schools Council Project on Statistical Education: 1980, *Statistics in Your World*, Foulsham Educational, Slough.

Shaughnessy, J. M.: 1991, 'Research in Probability and Statistics: Reflections and Directions', in D. Grouws (ed.), *Research on Teaching and Learning Mathematics*, in press.

Shulte, A.P. (ed.): 1981 *Teaching Statistics and Probability*, 1981 Yearbook of the National Council of Teachers of Mathematics, Reston, VA.

Tournaire, F. and S. Pulos: 1985, 'Proportional Reasoning: A Review of the Literature', *Educational Studies in Mathematics*, **16**, 181-199.

UCSMP Newsletter: 1987 (Spring), University of Chicago School Mathematics Project, Chicago.

Willoughby, S., C. Bereiter, P. Hilton, and J. Rubinstein: 1987, *Real Math Program*, Open Court Publishing Company, LaSalle, Illinois.

Wooton, W.: 1965, *SMSG: The Making of a Curriculum*, Yale University Press, New Haven, CN.

Andrew Ahlgren
American Association for the Advancement of Science
1333 H Street NW
Washington, DC 20005
On leave from University of Minnesota
USA

Joan Garfield
University of Minnesota
340 Appleby Hall
128 Pleasant St.
Minneapolis MN 55455
USA

Chapter 5

The Theoretical Nature of Probability in the Classroom

H. Steinbring

This chapter presents an epistemological analysis of the nature of stochastical knowledge. In particular, the mutual relationship between the elementary concept of probability (in its empirical form of relative frequency and in its theoretical form of Laplace's approach) and the basic idea of chance is demonstrated. An important consequence for teaching elementary probability is that there cannot be a logical and deductive course for introducing the basic concepts and then constructing the theory upon them; developing stochastic knowledge in the classroom has to take into account a holistic and systematic perspective. The concept of *task system* is elaborated as an appropriate curricular means for treating the theoretical nature of stochastic knowledge in the classroom.

1. Approaches towards Teaching

Teaching processes are intended to enable pupils to learn and understand mathematical concepts, methods, calculation techniques, and structural relationships. The first step is to simplify complicated mathematical problems, transforming them into an elementary form to facilitate learning processes. A teaching-learning path through the discipline's structure has to be found which is as easy to understand as possible. This means that teaching mathematics is aligned to carefully selected examples, and that the teacher skilfully attempts to establish new knowledge which has to be learnt on the basis of familiar things. Moreover, there is an effort to avoid ambiguity of mathematical concepts by means of clear characterizations, and to guide and control the process of learning by means of establishing a consistent structure of knowledge.

Teaching mathematical knowledge in school is bound to a context of social communication, within which the pupils are presented with a diversity of requirements relating to subject-specific activities, as well as relating to cooperation with classmates. When the

R. Kapadia and M. Borovcnik (eds.), Chance Encounters: Probability in Education, 135–167.

teacher organizes classroom activities, he/she will directly or indirectly refer to the subject matter of the discipline and to its structure. This helps to determine the mode of work in class: teacher exposition, teacher-pupil interaction, pupils' individual work, group work, project work, and so forth. Practising and repeating stages in the classroom are also closely connected with the demands of the subject. Teacher's activities aim to help pupils assimilate subject knowledge.

Structural approaches

Mathematical knowledge and the teacher's conception of it determines the teaching process to a very high degree. Thus, the educational approach of simplifying the subject matter and explaining it consistently is frequently identified with the logical structure of mathematics. We must, however, remember that a consistent presentation of mathematical knowledge does not necessarily yield the ideal structure for the teaching and learning of mathematics. Teaching and learning certainly cannot be characterized as a logically consistent method. As opposed to a one-sided and structurally oriented method of work, a broad scope of different activities of teaching and learning is required for mathematics education; this is also stated in the Cockcroft report (1982, §243):

> "Mathematics teaching at all levels should include opportunities for
>
> - exposition by the teacher;
> - discussion between teacher and pupils and between pupils themselves;
> - appropriate practical work;
> - consolidation and practice of fundamental skills and routines;
> - problem solving, including the application of mathematics to everyday situations;
> - investigational work.
>
> In setting out this list, we are aware that we are not saying anything which has not already been said many times and over many years. The list which we have given has appeared, by implication if not explicitly, in official reports, ... and the journals and publications of the professional mathematics associations. Yet we are aware that ... there are still many [classes] in which the mathematics teaching does not include even a majority of these elements."

Such exhortations have frequently been made and are basically accepted by mathematics educators. There remains, however, the question why these principles have little effect on real teaching. Good intentions alone do not seem to suffice, so the conditions for realizing different classroom activities must be inquired into.

It may be that classroom activities are closely connected with the conceptions teachers have of the *character of mathematical knowledge*. The epistemological status of mathematical knowledge in the classroom determines to a high degree the methods and activities which are deemed appropriate and suitable for learning mathematics. The unified character of mathematics and its logical structure are in stark contrast to the multitude and heterogeneity of classroom activities. Most teachers consider school mathematics and the curriculum as strictly fixed, firmly predetermined and structured hierarchically. From such a perspective, they organize their classroom activities according to the structure of their own mathematical knowledge. Hence, the teacher hands mathematics down to the pupil like a ready-made object in a simplified if piecemeal way (Steinbring, 1985).

Two teaching strategies fit this line. The teacher may orient himself/herself more towards questions of intuitiveness, of making the subject matter concrete and of illustrating it by means of examples. The intention is to present the subject matter to the pupil in an illustrative way, subdivided into small segments, and in a way which can be concretely represented. This method of teaching is to be found mainly at the primary and at the beginning of the secondary level.

Within the secondary level (Sekundarstufe I, grades 5 to 10), another method of teaching is developed in which the emphasis is on consistent mathematical connections. The mode of justifying and organizing the teaching and learning is aligned to the internal structure of the mathematics curriculum, which must necessarily be organized in a consistent way. This method of teaching does not mean that the logical structure of mathematics is immediately presented to be learnt, but that the focus and orientation of teaching is on bringing it out.

This is borne out by the dominant character of the question-and-answer pattern in direct teaching, the avoidance of experimental and practical work, as well as the almost pervasive three-step teaching pattern of presenting examples, developing mathematical rules or techniques for calculation and then practising these rules by means of further problems or applications (Dörfler and McLone, 1986, p.63).

The two teaching strategies, the intuitive-concrete and the structural method, both start from the assumption that mathematical knowledge and awareness is acquired in a linear, straightforward way. This means that the teacher must initially define basic concepts in order to be able to gradually introduce the further elements of knowledge into this framework. Conceptually clear and fixed foundations are required to ensure that mathematical knowledge can be learned according to this view. The basic assumption seems to be that the axiomatic structure already contains, at least in principle, the entire knowledge.

Winter (1983) writes about the consequences of this structural approach in teaching

> "The concern for clarity of concepts always runs the risk of degenerating into an obsession with precision. The latter is evident, among other things, 1) in a readiness to provide a lot of definitions (and only few theorems, these being even trivial) which kills motivation; 2) in a tendency to define things which are not systematically used at all; 3) in the obsession of using as many specific terms of the discipline as possible; and 4) in the concern to strive for definite, watertight formulations from the very outset." (p.196/197)

Classroom learning is directed towards the mathematical conclusions drawn from these foundations. This perception of school mathematics leads to the desire to have precise and comprehensive definitions from the beginning for as many concepts as possible - even for the insignificant ones. In due course, the teaching of mathematics runs the risk of aligning classroom activities within the teaching-learning process too directly to the logical mode of mathematics itself. This may quickly transform the process of teaching into a logical, consecutive procedure which excludes the ruptures and contradictions occasioned by the social and communicative context.

Freudenthal (1983, p.41) uses the example of the *reference set* to formulate a similar criticism with regard to the alleged need for a comprehensive set of basic concepts.

> "Curriculum developers and textbook authors apparently believe they offer teachers and pupils firm ground under their feet as soon as they restrict mathematical activities to a fixed reference set. This, they think, prevents possible surprises that might be disappointing. Fixed reference sets have been a source of confusion - in particular in probability (which we are obliged to anticipate here) - and have provided, rather than firm ground under their feet, a swamp of contradictions."

This reference set cannot be an invariable foundation for mathematical knowledge. On the contrary, in teaching, the reference set must be constructed and adapted according to the actual state of the knowledge in class.

Teaching within this narrow conception of the nature of school mathematics is not consonant with the genuine process of learning theoretical knowledge. In the structural method, knowledge is unequivocally defined beforehand both by the basic concepts and their representation and by the mathematical methods of inferring new knowledge. In contrast, processes of learning require the organization not only of the development of knowledge, but simultaneously of its modes of representation and of the opportunities for active engagement.

Processes in the mathematics classroom thus have to face the problem that there are no ready-made basic concepts from which, in principle, we can derive all mathematical knowledge. Teaching has to consider knowledge as dependent on the available means of representation and activity and on contexts. The knowledge to be taught has to be developed at the same time as the means of representing it, and the situations to become active with it. The meaning of mathematical knowledge is not immediately obvious; it is dependent on object-related representations, and on the means of working on these representations. This meaning is developed by actively dealing with objects and their specific epistemological nature; Freudenthal (1983) introduces the notion of *mental objects* to express that.

Beyond that, teaching must also take into consideration how mathematical knowledge can be made effective and meaningful in general education.

> "... scientific concepts always include general aspects, the theoretical objects of teaching are neither visible nor palpable. Their transmission therefore cannot be carried out like, say, throwing a ball. The teaching of theoretical concepts must take this into consideration and requires, consequently, a conception of the role of the activity, of the intuition, of the language, and of the logic of concept-development in mathematics instruction. Another aspect is related to the problem of general education and of its social importance, or its function in the development of personality." (Otte and Reiß, 1979, p.124).

2. The Theoretical Nature of Stochastic Knowledge

The tendency of strictly organizing teaching according to ideas about the character of mathematical knowledge also dominates stochastics instruction. The consequences are even worse than in other branches of mathematics education as stochastic knowledge has a specific theoretical character which will be discussed here. The role of this theoretical character of stochastic knowledge is considered for the teaching process and its organization. It implies that the basic concepts cannot completely define probability, but are conversely formed only while developing probability theory so that the teaching process must be organized differently. It is inappropriate to start from ready-made concepts, gradually adding further knowledge. Conversely, it is necessary to begin with meaningful situations which facilitate the development of concepts.

Approaches to teaching probability

The two main methods of teaching are again based on intuitiveness and on consistency. The intuitive approach focuses on experiments, ideal games of chance, and real situations together with their conditions. Frequencies, numerical proportions and symmetries are used to define probability. After this intuitive and experimental phase of teaching, there is a distinct progression to statistical methods and concepts.

Somewhat opposed to this intuitive mode of teaching is a method which places emphasis on the internal consistency of the concepts, and on the inner foundation of the stochastics curriculum. This means that the teacher introduces a number of basic concepts: outcome, result, event, elementary event, certain and impossible event, sample space, etc. These concepts are intended to provide a solid foundation for subsequent development. The specific intention is to give clear definitions of concepts, free of doubts and ambiguities. This method of teaching almost always uses the concept of classical Laplacean probability to underpin the mathematical structure of the conceptual connections.

Frequently, there is a temporal succession between these two basic types of teaching stochastics. The teacher begins, in the initial years, with concrete work. Later, the theory is developed consistently, loosely using the preliminary intuitions of the outset. However, there is no discussion about the conceptual link between the two approaches; the fact that either part is only a different perspective of the same probability concept is not discussed.

As a consequence, probability and statistics are separated in the classroom. From the view of probability theory based on the classical Laplacean approach, statistics runs the risk of becoming a collection of methodological recipes without theoretical connection; moreover, the Laplacean conception lacks reference to real situations which are not equally likely and thus lacks the full range of meaningful applications.

In spite of these different perspectives in the two basic types of teaching, both start from the idea that mathematical knowledge can be gradually enlarged and integrated into the overall structure of the curriculum by adding small elements in a piecemeal way. This is best illustrated by an analysis of the use of the basic ideas of *probability, randomness and chance experiments.*

Usually, one of the two possibilities, relative frequency or classical probability, is used as early as possible. The first approach is through ideas of stabilization based on the empirical law of large numbers, in order to define the concept of probability as the limit of relative frequencies and subsequently reducing the concept of equiprobability to that. Alternatively, the intention is to refer exclusively to the concept of classical probability by restrictions to symmetrical experiments. In both cases the objective is to have a clear and mathematically complete definition of probability as soon as possible to provide a firm grounding.

Before the concept of probability can be explicitly defined, however, the object of probability calculus, in this case the chance experiment and randomness, must be explained. Thus, explaining the concept of the chance experiment is the beginning of any stochastics instruction. Any experiment which cannot be predicted or determined, is dependent on chance. Experiments with several outcomes whose appearances cannot be forecast, are accordingly called chance experiments. Another definition of chance experiment is that it is simply an experiment in which the results are dependent on chance. A chance experiment is thus characterized negatively in comparison to a deterministic experiment or with reference to the probability concept which is still to be explained. Andelfinger (1983) uses the expression "self-referentially" for this latter kind of argument.

In order to illuminate the initially vague ideas of randomness and chance experiments, various concepts, such as result, outcome, event, sample space are introduced. There is no discussion, however, of what is to be understood by chance, how the concept of chance becomes meaningful in real life, and how chance can be specified mathematically (by means of the concept of stochastic independence, for instance) . The chance experiment is an introductory idea which is supported mathematically and linguistically in many ways, without developing its precise meaning and an underlying representation in the course of teaching.

These modes of teaching stochastics start from a vague notion of the object, that is of randomness and the chance experiment, whilst using a definite, mathematically precise definition of probability as soon as possible. These two basic concepts - *probability and randomness* - are no longer questioned later, on the contrary, they form the invariable basis from which mathematical conclusions are drawn in the course of teaching. This means that the teacher defines new concepts on this basis, performs certain calculations and attempts to solve problems by means of established methods and procedures.

Theoretical nature of probability

The two epistemological characterizations of probability by means of relative frequencies or by means of the classical definition can be interpreted as different ways of introducing and defining the same concept. The aim of defining basic concepts in a universal and absolute way leads to *vicious circles*. This is because what is to be defined must already be assumed for a universal definition. A well-known example is the concept of equiprobability, which becomes circular in the case of its universal interpretation: the definition of equiprobability is based on equal likelihood of possibilities.

As a basic concept, the concept of relative frequency is not free of such circularities, either. Relative frequencies, or the limit of relative frequencies, can be interpreted as probability only if they refer to a stochastic object such as a stochastic collective. The definition of such a collective, however, assumes that concepts like chance and independence, for instance, are used simultaneously. Here, too, conceptual circularities in the foundation of probability are evident, which means that concepts must be assumed which can only be developed and explained by means of the theory; a detailed analysis is in v. Harten and Steinbring (1984).

The circularity within the definition of the concept of probability can be more generally characterized as the mutual dependence of object (to which the probability refers) and concept (which is introduced on the basis of the object): one is able to speak of probability only if one knows the concept of the chance experiment and randomness; vice versa, chance can only be understood and mathematically specified after the concept of probability has been introduced.

The introduction of probability via relative frequencies while using concrete chance experiments is matched by the teaching method of developing concepts in an operationalized form. This means that concepts are not explicitly defined, but characterized by their use and application. Such an operationalized definition of the concept is expected to create a deeper and closer relationship with the pupils' subjective learning activities and learning processes.

Operationalizing a mathematical concept, however, cannot completely reveal its epistemological meaning. The instructions for use which are intended to define the concept might produce results which differ strongly from each other. For a chance experiment, the relative frequency obtained from a series of experiments, is usually not *identical* with the underlying probability, as different values occur in repetitions of the experiment, even if the conditions remain the same. The ways of operatively using a concept cannot characterize the concept's meaning; they have to be compared and related to find invariant epistemological aspects of the concept.

If the classical concept of probability is reduced to equiprobability in school, processes of abstraction and idealization play a substantial role in its development. This method abstracts from certain empirical properties; it assumes symmetry of given objects in order to introduce ideas and properties in a mathematically ideal way. This way of idealizing is used to introduce and describe concepts in other fields of mathematics instruction, too.

For stochastics, however, this procedure raises the problem of how the random character of chance situations can be modelled in such an ideal representation. The concept of probability is not simply a fixed quantitative relationship, as it results from ideal symmetry but this relationship is at the same time a 'random quantity' which varies in repeated experiments. Probability refers not only to an ideal structure, but has to reflect the real conditions of the random number generator considered. The concept of probability is thus not just a result of abstraction and idealization of empirical properties, it should also offer possibilities of modelling deviations from this ideal structure.

If taken absolutely, the frequentist and the classical approach to define the concept of probability result in circularities in their epistemological characterization. Their educational derivations in the shape of developing the concept by abstraction or developing the concept operatively are also tainted by one-sidedness and show that the respective opposite aspect is required in order to understand the concept completely. Teaching stochastics cannot simply be based on explicit definitions of probability and chance which are provided so comprehensively that they contain, in principle, all the mathematical consequences. Rather, the concept of probability and the concept of chance must be developed while being mutually related to each other. In either case, one must start with limited and relative definitions and exemplary explanations of randomness and probability, developing them step-by-step by means of applying them to one another and of explaining them.

The central idea of developing concepts in this fashion is now illustrated by the concepts of chance and probability. What might be a first understanding, a simple definition of chance? One could begin with the following idea: a sequence of Os and 1s is a random sequence if no pattern can be distinguished in the succession of numbers. On the basis of such an implicit idea of the character of chance, the stochastic thinking of children can be developed.

Such a definition, however, is neither explicit nor sufficiently clear-cut. For any sequence, it cannot be stated with certainty whether it is random or not. Are the figures presented only the beginning of the sequence, which will recur later? If two sequences are compared, which sequence is random, or 'more' random? What can this mean mathematically? Just think of pseudo-random numbers as created by computer-algorithms; even the digits of π behave like a random sequence though they are generated by an algorithm.

The consequence for the definition is that randomness and the linked question of whether a sequence is random, cannot be decided a priori once and for all; conversely, chance is defined according to the method available to test the random character. Hence, a sequence is judged to be random if it satisfies all known tests of randomness. This does not define the concept of randomness universally since chance can only be analyzed according to the known statistical methods. The definition of randomness is thus dependent on the developmental level of the stochastics' course (see v.Harten and Steinbring, 1983).

The circularities in defining the concepts of probability and randomness conflict with teachers' intentions to create a solid foundation of knowledge by precise definitions of the basic concepts. The epistemological character of stochastic knowledge and concepts thus inhibits teachers' preferences to start with explicit universal definitions. The basic concepts cannot be provided, once and for all; rather, they must be developed step-by-step within the process of teaching. Jahnke (1978) and Steinbring (1980) use the term *theoretical character of mathematical knowledge* to denote this dependency of the concept and its meaning on the theory's level of development. In this view, the concepts are not simply the foundation for the theory, but conversely, it is only the theory which explains and develops the meaning of fundamental mathematical concepts.

Teaching processes in stochastics thus cannot be aligned only to the mathematical structure of the concepts, they must shape the development of the concepts in accordance with the epistemological status of stochastic knowledge. It is necessary to create an educational framework in which the mutual relationship between probability and randomness, as well as between the various forms of representing probability allow for concepts and theory to unfold in unison. The teaching process must address the paradox that clarification and an understanding of the fundamental concepts can only be achieved by the fully developed theory in the further course of teaching. An important condition necessary to meet this requirement in the classroom is to avoid developing stochastics exclusively by means of relative frequency or via classical symmetry.

Objects, signs and concepts

The common method in the classroom of making calculations and solving problems on the basis of seemingly solid and final definitions of chance and probability is not sufficient even for teaching elementary probability. The connections with combinatorics or Boolean algebra are internal means of justifying stochastic concepts, stressing the aspect of mathematical signs and calculations, but at the same time neglecting stochastic model representations and interpretations of real situations of chance. This exclusive emphasis on mathematical formulae and calculation techniques hinders stochastic interpretation and thinking. Randomness is not automatically contained in the mathematical signs: beyond those, stochastics must be interpreted according to its content; this requires a stochastic situation as object to which these models refer.

The requirement to consider concrete random events and situations is not the same as stating that the empirical situation itself already provides the comprehensive meaning and explanation of stochastics. Randomness is not simply an empirical property of a given real situation, it is simultaneously a theoretical concept which must be developed, refined, and analyzed.

Neither the empirical situation by itself (the object or the problem), nor the mathematical sign by itself (the model, the calculus) can express what randomness and probability mean. Steinbring (1984, p.83) argues that both an autonomous object and a model or calculus, and the relations between them are necessary to understand and develop chance and probability.

The situation is illustrated by a triangle of object, sign and concept. We will show how this diagram of relations in fig.1 describes the particular character of mathematical knowledge and gives orientations for organizing learning processes in the classroom. First, however, we explain how the specific difficulties encountered in stochastic thinking make the theoretical character of knowledge inevitable.

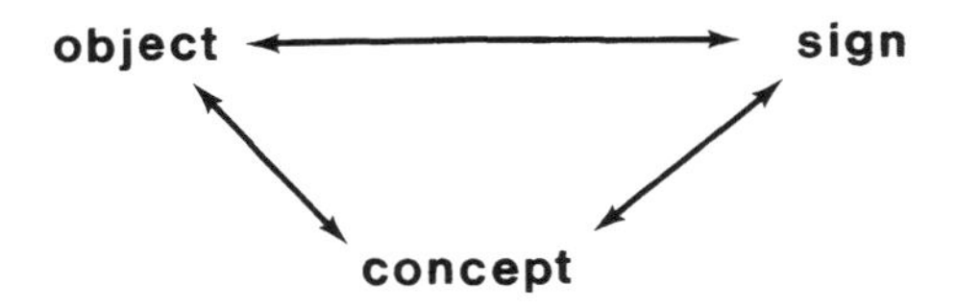

Fig.1: A triangle of relations between object, sign and concept

The previous epistemological analysis of the basic concepts demonstrates the necessity for a *dynamic* perspective on knowledge development. Knowledge cannot be determined completely on the basis of invariable concepts. Knowledge and its meaning has to be developed and this requires a flexible interpretation of these basic concepts. Such a dynamic perspective on knowledge is not only appropriate with respect to the epistemology, but also with respect to the student; the acquisition of mathematical knowledge can only be understood as a developmental process. It is a question of the relation between the student's pre-experiences and subjective notions and the mathematical structures and concepts. Such a broad subjective perspective for teaching probability should not be confused with the conception of *subjectivist* probability; but in a very elementary way important aspects of subjectivist intuitions are contained in this approach.

The epistemological analysis of knowledge is very helpful for a better understanding of the relation between subjective intuitions and objective random situations and their mathematical models. A dynamic perspective on knowledge helps to soften the strict alternative between subjective intuition and objective structure. If one regards the relational triangle from a developmental perspective, the subjective aspects of the learning process become more apparent. Beginning with very personal judgements about the given random situation, comparing the empirical situation with their intended models will hopefully lead to generalizations, more precise characterizations, and deeper insights.

The many paradoxes of probability theory illustrate the difficult relation between subjective intuitions and objective structures of probability. Paradoxes often show a contradiction between modelling and objects with respect to mathematical calculation and intuitive representation. These contradictions between calculation and representation result in part from a subjectively perceived difference between the mathematical result on the one hand and the intuitive representation of the objective situation on the other.

The difference between object and mathematical model cannot simply be regarded as an error or as a defect to be corrected; the difference between object and model is a matter of principle in stochastics. Model and object can never be made completely congruent; one can look for models which are adapted to the real situation in a better way. The implicit assumptions of the model must be checked as to whether they are justified for the real situation; there is also the question of which of the situation's concrete features should be additionally taken into consideration if the model does not fit satisfactorily.

The discrepancy between a real situation and its model holds for all applied mathematics. For probability, however, this remaining basic difference is a genuine object of theoretical analysis. Stochastics really is a mathematical theory about errors, deviations and interrelationships between an idealized forecast and real fluctuating behavior. This difference and this interrelationship between object and model together propel the development of stochastics. In this respect, in any stochastic analysis there is always a certain amount of subjective judgment involved, which cannot be completely eliminated though one wants precise information about its extent.

3. Didactic Means to Respect the Theoretical Nature of Probability

The previous considerations have shown that this relational triangle of object, sign and concept is a useful tool to characterize stochastic knowledge avoiding a one-sided emphasis either on mathematical calculations or on empirical-concrete aspects. This section explores how the relational triangle develops educational orientations which enable the teacher to bring the theoretical character of stochastics to the open in the classroom. Hence, it analyzes educational means which permit a description of the actual process of developing mathematical knowledge.

Interrelations between mathematics and exemplary applications

The significance of the relationship between empirical stochastic situations and modelling can also be traced in the historical development of probability theory. Loève (1978) shows that the development of the concept of probability has been advanced in particular by the tension between empirical application and mathematical description. These concepts were thus established as a result, or as descriptions of this relationship, at the same time serving to analyse this relationship and to develop it further. Steinbring (1980) applies the principle of complementarity to describe this dynamic relation between a theory and its applications.

The principle means that stochastic concepts must be simultaneously understood as elements of mathematics and as objects related to a context of application. Neither a mathematical definition nor an interpretation of exemplary relationships and applications alone establish the probability concept; it is the dual interplay between both sides that counts. In the sense of this *complementarity*, the concept of probability should be understood as a relationship between mathematical description and exemplary cases of application. This principle prevents conceptual reduction, be it to mathematical calculations, or to empirical intuitions; it excludes philosophically one-sided conceptions of probability like a pure objectivist or a pure subjectivist interpretation. Dinges (1981) also opposes such a conceptual reduction:

> "Stochastics in school must organically combine ideas from different philosophical traditions, in particular
> - stochastics as the mathematics of mass phenomena
> - stochastics as the logic of uncertainty
> - statistics as the technique which transforms data into insights
> - stochastics as decision theory." (p.51)

Like probability, the concept of randomness bears these complementary elements. Random situations and chance experiments are concrete applications or real situations of chance like random number generators, but theoretical methods and procedures are required to describe and analyze randomness with regard to its mathematical relationships. Even a concept as elementary as equiprobability simultaneously needs the ideal mathematical interpretation of equally distributed in the arithmetical sense, and the empirical representation of statistically equally distributed, as it emerges in concrete experiments.

Equidistribution as a model does not mean that actual results should exactly be equally distributed; if this really happens it is a sufficient reason to mistrust the experiment. The model of exact equidistribution is required as a reference to decide whether observed frequencies can be judged to stem from such a uniform distribution, or if deviations are too large for such a claim. In the case of a loaded die, the usual variability of equidistribution is essential to give an idea about the extent of falsification which is possible on the basis of the observed values. This example shows the complementary nature of stochastic concepts. The concept of equidistribution establishes a relationship between a real stochastic situation and its modelling, a relationship between an empirical, statistical, and an ideal, arithmetical equal distribution. The concept of 'equally distributed' really has this double meaning and neither of these two aspects may be ignored.

In an analogous way, the concept of the expected value mediates between an empirical situation and its mathematical modelling. The notions of mean values and values of dispersion, which characterize a distribution are related to the conceptual neighbourhood of expected values. These latter are ideal values which must always be considered in relation to empirical values and concrete data; they serve as an ideal point of reference for empirical random quantities.

Bernoulli's law of large numbers was the first mathematical theorem on the connection between relative frequency and classical probability. It reveals, in its logical structure,

the complementary relationship between a stochastic situation and its mathematical model; at the same time, it serves to analyze and specify this relationship. It refers to an experiment which has two outcomes 0 and 1. The theorem states:
If p is the probability of a 1, with $0<p<1$, and H_n/n is the theoretical relative frequency of 1 in n independent trials, then it holds

$$\forall \varepsilon>0 \quad \forall \eta>0 \quad \exists n_0: \quad P(|H_n/n - p| < \varepsilon) \geq 1-\eta \qquad \forall n \geq n_0$$

The reflexive statement in Bernoulli's theorem says that there is a very great probability that relative frequencies in a stochastic experiment will come as close to the probability as desired if the number of trials increases. Altogether we have three variable quantities in the theorem; first, the precision of the statement considered which is measured by ε; second, the certainty with which the statement holds which is measured by η; and third, the number of trials necessary which is given by n_0. These three parameters are mutually dependent on one another, it is possible to fix two and then to calculate the third.

The need to use probability in order to even formulate how relative frequencies 'converge' to the underlying probability is an indicator of the circularity of the concept's definition and of the complementarity between the empirical situation and its mathematical model. A detailed educational analysis of Bernoulli's theorem is given in Biehler and Steinbring (1982) or in v. Harten and Steinbring (1984).

All stochastic concepts examine the mutual relationship between object and mathematical model, they basically emerge in this connection, serving to analyze and explore this difficult linkage further. The concept of probability thus cannot be defined once and for all in a definite form, nor does the concept of chance represent a universal basis which serves to derive all further mathematical statements by logic. The meaning of the concept and its relationship to applications emerge only from the development of the theory and the refinement and elaboration of this relationship between object and model. This dialectics of empirical content-specific and mathematical model-specific aspects establishes what we call the 'theoretical character' of stochastic knowledge.

Means of representation and activities

The idea of complementarity prohibits the hierarchical method of teaching by beginning with content-related representations of applications, and then passing on to the symbols, formulae and mathematical modelling which represent an abstraction of the real situation. Usually, the empirical situation is intended to facilitate modelling, which is then followed by generalization. In principle, the interrelationship between object and model must always be kept in mind throughout the teaching process. The dialectics of object and model requires a tempered approach.

At the outset there should be a situation in which the relationship between intuitive representations and mathematical modelling statements is a crucial element which promotes development. Only those stochastic situations showing a multitude of different aspects are appropriate to stimulate pupils to investigate ideas. The need for subjective experiences and intuitions can be met by such problem situations which can then be examined and, if necessary, modified by comparison with a stochastic model, developed for the particular random situation.

As the process of learning begins with relatively complex situations, certain assumptions are made implicitly at first. These are assumptions about the stochastic character of the situation such as equiprobability, or the independence of experiments. In the course of teaching, these assumptions have to be dealt with, questioning the appropriateness of the modelling implicitly assumed at first, and improving and changing it if necessary. Analyzing and developing the implicit assumptions are vital elements of concept development which is governed by the tension between concrete stochastic events and their mathematical modelling.

If we agree that the mutual relationship between object and model in stochastics cannot be ignored, the question remains as to how to develop this relationship mathematically. Historically, a multitude of conceptual means of description and presentation have been developed to characterize the relationship between concrete individual situations and their stochastic modelling. Closer analysis reveals two categories of representation to exemplify the tension between a situation and its stochastic model.

The distribution category contains all the diagrams relating to the processing of frequencies. Relevant for the classroom are tables, tally sheets, pie charts, stem-and-leaf displays, column graphs, histograms, polygons, etc. The other category includes different types of tree diagrams which mainly deal with probability calculus. For school, the following forms are relevant: tables, complete and reduced tree diagrams, 'infinite' tree diagrams, Pascal's triangle and the probability abacus.

These types of representation are used as an auxiliary means to process data, or as a method leading to the mathematical formula for calculation. With regard to the relationship between a situation and its model, however, it is appropriate to consider distributions and tree diagrams as being complementary to each other. This means, for instance, that the concept of distribution can also be used to calculate and represent probabilities; conversely, Pascal's triangle from the tree diagram category can be used to explain the binomial distributions.

Comparing different representations leads to new insights into the relations between a concrete situation and its stochastic modelling. Each representation treats this relationship in a specific way; each means of representation reflects the complementarity of stochastic concepts. The different modes of application which these representations permit, are an expression of these complementary aspects. They enable us to build a broad spectrum of relations between a concrete situation and its mathematical model.

Thus, the tree diagram may be interpreted both as a flow chart for carrying out a concrete experiment, and as a characterization of all the possibilities occurring in the shape of a model. Hence, the tree diagram can be used to describe specific experiments; it can also be used as a scenario for a large number of idealized experiments; it may also serve to derive the path rules (addition and multiplication rules) and to establish stochastic models like the binomial distribution.

The different variants of the concept of distribution can be used to represent, order, and relate data. It is possible to make absolute or relative comparisons, either by areas or lines. Different scales or certain transformations such as square or logarithmic functions may reveal structures in the data. Distributions characterize simultaneously the overall system of the observed quantities and show specific salient features like symmetry, means, variances, etc. The educational role of graphical representations is dealt with in Biehler (1985). These specific characteristics of distributions and the analysis of random situations enable us to compare empirically observed distributions and their possible mathematical modelling.

For mathematics instruction, the means of representation and activity are of fundamental importance as they allow the development of concepts and the formation of new knowledge. The different modes permit the first generalization beyond the present level of knowledge. At the same time, for their concrete shape, they are an expression of a limited point of view, i.e. they indicate a potential framework of the development and of possible steps. The different modes of representation are an expression of the fact that the acquisition of knowledge, on the one hand, proceeds in stages and leaps and bounds, while the interrelationship between these stages is at the same time potentially and tentatively established. This conflicts with a conception of the development of science in which *addition* of new elements of knowledge is possible on the basis of a given structure.

Thus, the means of representation and activity is a 'mediator' between the subjective individual growth of knowledge and the curricular, epistemological development of knowledge. They do not only describe a potential curricular process of development but they also offer opportunities of connecting pupils' individual learning to this curricular process.

The relational triangle of object, sign and concept, the idea of complementarity, and the discussion of representation and activity prove the following distinctions useful: knowledge has to be distinguished from the forms to represent it; knowledge has to be distinguished from applications or interpretations of it. Hence, the teaching process requires more than simply organizing the development of knowledge. At the same time teaching has to organize the forms of representing this knowledge and the opportunities to become active with it.

Forms of activity in the classroom might not directly refer to stochastic knowledge but they must respect the complex structure of object, sign and concept. Therefore, in preparing and conducting their teaching, teachers have to search for suitable means of activity and representation. For theoretical knowledge is not directly accessible; to acquire it, one has to use mediating concepts. The frequent use of means of activity and representation in teaching is an indicator of the theoretical character of knowledge. The diagram in fig.2 illustrates the underlying relationships in an effort to find a balance between the knowledge to be learnt and the activities in the classroom which are to serve that purpose.

Our analysis of stochastic knowledge reveals that this knowledge is theoretical and complex; its logically structured representation cannot claim to cover the whole meaning and explanatory power of stochastics. This conception of the character of stochastics knowledge and the development of means of representation and different modes of use and activities form the basis required to achieve successful learning processes. Logically structured and hierarchical exposition of knowledge is not sufficient.

teaching requirements orientations, methods	<-------------------->	school mathematical knowledge
↕		↕
methods and activities	<----------->	signs and representations
make experiments	<------>	random generators/ random situations
simulate	<------>	use a model of the real situation
count/ calculate	<------>	numbers, data
evaluate formulae	<------>	numbers, signs
set up formulae	<------>	signs, diagrams
.		.
.	<------>	.

Fig.2: Balance between activities and representations

4. On the Didactic Organization of Teaching Processes

Practical teaching requires a variety of concrete and effective aids and suggestions for *possible* approaches in the classroom. As has been discussed, the theoretical character of probability should bear a strong influence on educational approaches in the classroom. On the one hand, the demands of the subject matter require the teacher to develop special ways of teaching and planning; on the other hand, the concrete conditions in classroom and the crucial role of the teacher within the learning process have to be respected.

The role of teachers

Stochastics is a topic which is new for most teachers. Attempts to introduce new subject matter in mathematics instruction and to make concrete suggestions for implementation in the classroom are often ambiguous about the teacher's role. Above all, the teacher is considered to be a deliverer of the mathematical knowledge. This leads to an effort to provide the new materials in a detailed and ready-made way to ensure that it can be immediately integrated into teaching. Too often, educators and researchers as well as teachers work on the basis of such a limited understanding. Teachers implicitly expect in-service training to provide immediately effective material, while this attitude induces educators and researchers to develop suggestions which can be directly applied in the classroom.

There are various suggestions which stress the internal mathematical consistency and logical hierarchy of knowledge which belong in this realm of so called 'teacher proof' curricula for stochastics instruction. One approach is mathematically based on Boolean algebra and makes an exaggerated use of set theory. The other, more frequent approach stresses Laplacean probability and combinatorial aspects, as these facilitate a mathematically tidy presentation of the subject matter. The elegance and structure of these proposals suggest to the teacher that implementation of the curriculum will create no problems in the classroom.

This approach also includes courses in which concepts are developed by means of a universal concrete model. For instance, Freudenthal (1975) fosters his course entirely around the urn model; Bentz and Palm (1980) use the wheel of fortune instead. A concrete stochastic model permits special points of reference, explanations, and modes of work in the classroom. This model may serve to disclose the mathematical structure and logical order and, moreover, as a medium to model real applications. However, if used exclusively, this approach bears the risk that the chosen model itself becomes the sole object of teaching for the conceptual basis of stochastics. This universal model would then be overladen with its different roles, namely standing for the concept, being a concrete model for a situation, and being a prototype for the modelling process as a whole. Furthermore, this concrete model might shift attention to calculations so that questions of stochastic interpretation, application, and evaluation are in the background.

The intention of preparing teaching material in a perfectly structured and ready made way to by-pass teachers, has failed. The emerging problems of comprehension of students require an active intervention by the teacher and corresponding adjustments of the teaching material. The teacher must actively interfere with this process; he/she must shape, modify, organize and evaluate it. Classroom problems with 'teacher-proof' curricula have led to encourage teachers to develop their own material. Such a view of the teacher's role was also, involuntarily fostered from the perspective of genetically-developmental teaching. This conception of teaching sees the teacher as an active decision--maker and organizer of classroom activities. Only the teacher is familiar with the prior knowledge and abilities of his/her pupils and the conditions in the classroom. However, it is unrealistic to expect teachers to develop materials independently from one another.

Both extreme views on the teacher's role do not improve actual teaching. Neither is the teacher irrelevant with regard to the subject matter taught, nor is it sensible to view the teacher as a curriculum developer. The demands and necessities of practical teaching must be taken into consideration when developing materials and teaching suggestions; it is evident, however, that this consideration for conditions relevant in the classroom means that it is not possible to make suggestions for teaching which by-pass the teacher. Teachers have to acquire special knowledge to professionalize their daily practice. This professional knowledge of teachers must always reflect the mutual relationship between classroom requirements on the one hand, and discipline-specific and educational connections on the other.

For the teacher it is neither sufficient to know the systematic structure of stochastics nor is it sufficient to understand stochastics merely as subject matter to be taught to the pupils. The latter includes knowing the pupil's learning situation and knowing how to facilitate the presentation of the subject matter. Rather, the teacher must develop an idea of the relationship between the structure of the discipline and the pupils' acquisition of knowledge. In a pragmatic sense, this idea encompasses elements which are theoretical and systematic as well as including examples and experiences which are of practical effect and importance in the classroom. The special character of professional teachers' knowledge is formed by this 'mixture' of theoretical and practical elements, which means that this knowledge is neither final nor only theoretically based.

At the same time, it means that this knowledge is neither exclusively pragmatic nor immediately effective. The teacher has to relate the different social, pedagogical and psychological elements, and elements of the subject matter to one another within the teaching/learning process. In doing so, however, he/she has to keep in mind that the classroom process itself is subject to pragmatic decisions with practical effects which cannot be organized according to a systematic modelling of the discipline in the sense of a scientific justification. Classroom approaches will be effective only if they give due weight to the essential role of teachers and their special professional knowledge.

The role of task systems

In previous sections, we have shown that the character of stochastics is necessarily theoretical. Hence, suggestions for teaching must be developed within the context of the teacher's professional activity and teaching practice. These suggestions, however, cannot anticipate the practice of teaching in all its details, thus making the teacher's role superfluous. What kind of proposals meet this difficult requirement of containing simultaneously theoretical and systematic elements, and orientations effective for practice? What sort of teaching materials support the teachers' active role, whilst permitting alternatives and deviations without limiting these materials to a one-sided viewpoint?

The *mathematical task* plays an important part both in the classroom and in the teacher's understanding of mathematics instruction. Teachers plan and organize their teaching in line with mathematical tasks. Problems constitute a link between ideas which are effective in the classroom and the practical requirements; the idea and conception of a mathematical task serves to connect theoretical orientations and activities with practical effectiveness (Bromme and Juhl, 1985).

Tasks have a multitude of functions in teaching. They are the smallest units of classroom activity, which enable the teacher to relate the social-communicative and the subject-specific content of mathematics instruction to each other. Mathematics teachers often think and talk about teaching in the form of tasks; tasks are in some way their 'concepts'. However, they use these concepts pragmatically as practitioners; these concepts are highly interwoven with the local requirements, and lose their meaning without this context. Theoretical ideas of these concepts are relatively underdeveloped; the appropriate use of tasks requires at first many practical experiences and this can be observed in what is called a 'good teacher'.

Mathematical tasks which make sense never occur separately. They are always connected to a larger problem context which conveys meaning and coherence to a certain degree. It is in this way that serious mathematical tasks are distinct from puzzles and braintwisters. The context of tasks in the classroom consists mainly of other tasks. An optimal form of presenting several tasks together in the classroom is neither a number of routine exercises nor a haphazard mixture which are vaguely linked to each other. Hence it is important to develop a concept of a system of tasks which is between these extremes.

The most important aspect is to find a *structure of relationships* for tasks which belong together, in short, to interpret tasks as variables. This heuristic image of 'tasks as variables' is a good first approach to systems of tasks. Besides the mathematical context, there are variations of pedagogy, teaching method, means of representation and activity; possibilities of variations of the most diverse kind which may even contradict each other, thus leading to compromises in practical teaching.

What are the features which characterize task systems? What links individual tasks together to form a *system*? On the one hand, these tasks should relate to one *common* object - even if they do so from different perspectives. On the other hand, the tasks should be *analogous* to one another. It is evident that this is not a precise description of the concept of task system, for only working with tasks and solving tasks will show that they refer to the same object and are analogous to each other.

Task systems thus are 'open' to a certain degree; their system character is not simply given a priori, but constituted by the application, that is by using them and working with them. The concept of analogy is of special importance for constructing and describing task systems. Pólya (1967) has advocated the search for analogous problems as a strategy to solve problems. He explains that the crucial element of analogy lies in the similarity of certain types of relationships.

> "Analogy has something to do with 'similarity'. The concept of similarity is vague in the sense of everyday language, and in geometry it is more restricted than the concept of analogy. Similar things agree in some aspect, analogous things agree in certain relationships between their respective parts." (p.52)

In order to construct a task analogous to one which has been presented, it is necessary to develop an overview from which this similarity of the type of relationships becomes visible. It is necessary to reach a point of vantage from which it is evident how different types of relations between tasks can be transformed into one another. On the one hand, analogous tasks thus agree in a certain relationship between their respective parts, on the other hand, it is clear that there will always remain a difference between them as well. Both are important characteristics of the systems concept: agreement of the individual parts under a global aspect and distinction between these parts in their local aspects.

Two important strategies to construct analogous tasks are varying the problem's constants and change of the model. We will discuss them with the help of the 'birthday problem': 'In a room, a party of n persons has gathered. What is the probability of at least two persons having their birthday on the same day of the year?' Instead of asking for the birthday, one could ask for the birth-month, or the birth-week. Such variations are simplifications of the given problem. The birthweek problem, for instance, can be simulated much faster and more clearly than the original problem. However, part of the surprise at the high probability of repeated birthdays is lost because the coincidence probability is neither related linearly to the number of birth categories nor to the number of persons.

Varying the problem's constants may lead to an easy simulation of it. Simulation is an essential method in stochastics as it not only provides a method of solution, but its actual performance already requires a first model. Concrete drawings from an urn, throwing one or several dice, or coding and noting suitable random numbers from a table of random numbers are experiments which may lead to tree diagrams and a general stochastic model. Therefore, easy access to a feasible simulation of a problem does not only give a numerical result but also a conceptual understanding of the situation.

In the birthday problem, the path does not lead immediately from the simulation to the solution. This is due to the condition of 'at least two persons', a logical peculiarity which is frequently best treated by means of negation. In this case, however, the simulation may also serve to develop a model for the task and the problem posed.

Varying the constants may permit the construction of analogous tasks which open new paths to solution. Is this always the case with this form? Even in the case of the birthday problem, the seemingly harmless variations sketched may rapidly lead to important differences. For instance, the fact that months do not have an equal number of days, makes the assumption of equal probability dubious; it is less probable for a person to have her birthday in February than in August. Changing the wording from 'at least 2 persons' to 'at least 3 persons' leads to new difficulties. This variation goes beyond the initial model, it implies a *change of model*. A further complication is that the calculations frequently become much more difficult.

Such a divergence in the model is mostly an expression for an explanation of certain aspects of content, e.g. stochastic assumptions which suddenly become relevant for the model: these are mainly assumptions with regard to the independence of events like drawing with replacement, or assumptions such as equal probability.

For simple variations of the constants, the original model remains unchanged. Changes of the model in tasks can perhaps be integrated in a hierarchically superior model. At first, however, the tasks are mostly only analogous by way of a common problem object. Changes of the model, however, do not only occur in the case of extreme modifications. For instance, the following phrasing of the birthday problem, which might lead to a false intuitive perception of the situation, brings about an interesting change of the model: 'In a room, n persons are present besides me. What is the probability of at least one person having my birthday?'

These two forms of analogy, i.e. variation and change of the model, also represent a certain type of requirement asked of the learner: tasks within a system which are analogous because of a simple variation of constants, primarily serve purposes of developing and consolidating skills. In the case of a change of the model between analogous tasks, however, the emphasis is on aspects of the mathematical concept and of its development.

Task systems give an opportunity to organize forms of representing knowledge and activities of learning in a way which relates them better to one another. This gives due consideration to the theoretical character of knowledge. Besides these epistemological features of task systems, the relation to practice is important; educational differentiation is a case in point. A system should also contain tasks which provide access to stochastics for pupils with varying aptitudes and inclinations. Tasks within a system may be easy or difficult not only according to the subject matter, but also with respect to teaching. If we consider an entire system of tasks rather than individual tasks, the question whether a task is easy or difficult must be replaced by systematic questions like:

- What is the task's function within the overall system?
- How does its status change if it is removed to another position?
- Can the task be possibly replaced by another one?
- How does a task relate to certain other tasks of a system?
- Which are the different paths through a task system for pupils?
- To what extent are these paths equivalent?

There is also a large repertoire of different types of activities in stochastics: organizing experiments and simulations; obtaining and processing data; measuring, evaluating, and deciding; modelling using formulae and diagrams; these activities offer extra scope for variation of task systems. For a well organized system of tasks to be useful in the classroom, the variations relating to the subject matter and to the activities of learning must supplement and support each other. Such a task system may be used in a flexible way so that the combination of subject matter aspects and appropriate activities of learning offers opportunities for differentiation in the intended learning process.

5. Discussion of an Exemplary Task

As an illustration, parts of a stochastical task system for the initial teaching of probability are presented. The whole system together with a detailed analysis is contained in material for teachers to support their preparation for teaching (Brönstrup et al., 1986). Apart from the tasks and their solution, the material includes hints for the teacher concerning the didactic use of the tasks and their possible variations which are given parallel to the tasks. The rationale behind the development of these materials is discussed in Steinbring (1989). The following task is from v. Harten (1985, pp.61-63).

Didactic framework of the task

Fig.3 shows a board for a game. To play the game you need two dice. Choose a number from 1 to 12 and put your playing piece on the starting field with this number. Now two dice are thrown and the pips are added. If this sum equals your number, then your playing piece moves up one position. Whoever reaches the finish first, has won.

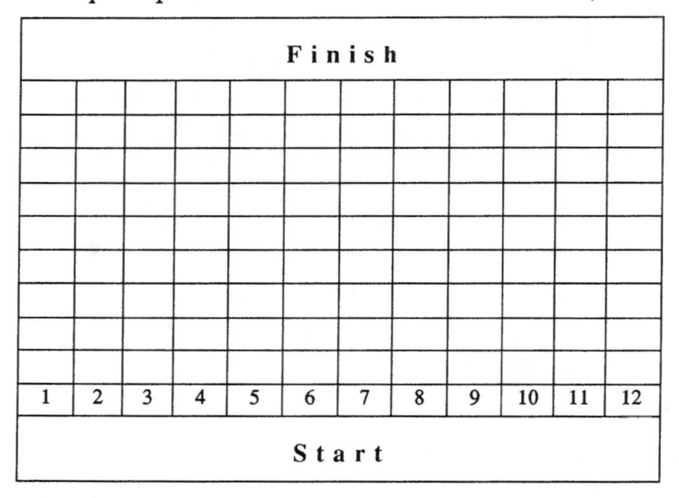

Fig.3: Board for the 'two dice game'

(a) For each throw, denote the sum of pips in the following table:

Sum of pips	Number of throws with this sum
1	
2	
3	
4	
5	
6	
7	
8	
9	
10	
11	
more than 12	

Total number of throws: ___________

Fig.4: Tally sheet for the frequency of the various sums

(b) Would you choose the 1 in this game? - Explain your answer!

(c) Which number or which numbers are good choices in your opinion? - Explain your answer!

At the beginning, the class might speculate and discuss experiences of winning strategies in chance games. Having played the game several times, these ideas can be checked. The material offers not only opportunities for statistical experiments and the calculation of relative frequencies as estimates for the probabilities, it also offers activities and tasks to explore the possible combinations of two dice in order to calculate the Laplacean probability of the various sums of pips. In an elementary way, the two conceptual aspects of probability are thus related to one another within this task.

There are hints for the teacher to use, vary and develop such tasks in the framework of a didactic task system. These hints link the different levels of professional knowledge. They have been revised after several uses in class.

Pupils' previous knowledge. Specific previous knowledge is not necessary but prior experience of graphical representations is useful.

Concepts involved. This task serves as an introduction to randomness and strategy. It introduces a more difficult random mechanism, the throwing of two dice. A more detailed analysis of this mechanism could be done with a subsequent task from the original system.

Methodical hints. This game should be played in small groups or by two pupils. Problems in understanding the rules of the game are rare; most pupils are familiar with this kind of game. Pupils in the groups may be given different duties: to throw the dice, to note the results, etc.

Pedagogical and technical notes. The tally sheet of this task should be kept by the pupils. It may be the first example of a distribution of results which are not equally probable. Other forms of representation may be introduced such as bar charts or histograms. A comparison with the table of a previous, similar task of the original system (if this task has been dealt with in the class) could be useful. One might use differently coloured dice.

Duration. This task can be carried out in one lesson. Experiences with work in various classes are: 5-10 minutes for explaining the game and the working sheets, 20 minutes for working in groups, and 10 minutes for the discussion of results with the whole class. How long pupils actually work in groups depends on how often they need to play the game. If a pupil chooses the favourable number 7, he/she needs an average of about 60 throws to reach the goal, deviations of more than 20 throws being rather unlikely. Numbers between 5 and 9 require not more than 90 throws on average.

Variations of the task. There are dice in the shape of a dodecahedron, with numbers from 1 to 12; to use such a die could give rise to interesting comparisons to the previous game. Three dice may be thrown and a board may be used with starting numbers from 1 to 18; this game is much more complex and time-consuming.

Classroom observations

This exemplary presentation of a problem belonging to the system is now supplemented and illustrated by observations from a classroom. In the course of their lessons on elementary probability, 6th graders repeatedly played a game of two dice in pairs during the previous lesson, noting their results on the tally sheet provided. In the lesson presented here which yielded the observation results, analysis of the game begins, and a simple theoretical explanation is sought for the unequal probability of winning this game with various starting numbers.

Most of the students recognized that the '1' never occurred as the sum of two dice; it really cannot occur, as they were quick to explain. They also noticed that numbers bigger than 12 are impossible. The sums 6 to 8 were obtained most frequently as they could see from their tally sheet. What is the reason for this? How can this be theoretically explained? This is the subject of the present lesson. Subsequent to a brief recall and a reminder of the experiment from the previous lesson, the teacher hands out a new worksheet to the students. They are to work on the following problem:

Problem: The lower table contains the various possibilities from which the sums of pips 2, 3 and 4 are composed. Please complete the table!

2	1+1		
3	1+2	2+1	
4	2+2	3+1	1+3
5			
6			
7			
8			
9			
10			
11			
12			

Fig.5: Tally sheet for the possibilities of the various sums

During the minutes which follow, the students work individually, or together with their neighbour. One of the students who has solved the problem quickly, is allowed to write his results onto a prepared overhead transparency. Gradually, almost all the students finish their work. Now the discussion of the results begins as a student-teacher dialogue. The possibilities are read aloud by students for each sum of pips and compared to the results on the transparency.

T.: What about 5? Isn't 2+3 the same as 3+2?... Frank?

S.: No, 'cause I can turn this around, and then it is something different!

T.: Let's pass to the 6. What possibilities do we have here? Jasmin?

S.: 1+5, 5+1, 2+4, 4+2, and 3+3.

S.: I've got something different!

T.: Yes, Nicole, what have you got?

S.: 3+3, 2+4, 4+2, 1+5, and 5+1.

T.: Only the order is different here!... But, what about 3+3? Don't we have to count 2 possibilities for this, too, just as for 1+5 and 5+1?

S.: If I turn the 3+3 around, I'll get the same anyway. This makes no sense, does it?

S.: For 2+4, you take again the opposite, that is 4+2, while you won't get anything new for 3+3. Or you write it down in mirror script, but that makes no sense.

T.: Just remember your game, the two dice! Yes, Christian?

S.: I'll mark the two dice. For instance, I'll mark one with A and the other with B, and you can have a 1 on one die, and a 5 on the other. Or the other way around, the die with B has a 1, and the die with A has a 5.

T.: Very good. And what about 3+3, Peter?

S.: This occurs only once. For 2+4 and 4+2 are two possibilities but with 3+3 there is only one.

These episodes give rise to many considerations and analyses. This brief section is merely intended to illustrate in an exemplary way the special treatment of the present task in the classroom. The students are very well able to establish all the possible combinations of two numbers for the various sums. The teacher's question why the sum 3+3 cannot be counted twice like all the other combinations is met with an explanation by reference to the concrete dice which is convincing within the context of the given task. One of the students imagines marked (i.e. distinguishable) dice and is thus able to give a pertinent justification for two distinctive combinations and for a single combination.

This brief episode covered only the beginnings of the classical concept of probability in the shape of 'all possible combinations of the sum of pips of two dice' which is later formalized by the rule 'number of all favourable events'. Nevertheless, it reveals the fundamental didactic principle: processes of understanding and developing concepts in the framework of a system of problems occur in interrelated work between simple models (symbols, operations, etc.) and contexts (applications, experiments). The students seem to deal with numbers, signs, sums etc.; their work, however, is organized with the previous game in mind, thus being enriched with possibilities of concrete justifications and meanings, as is shown by one of the students referring to two marked dice.

Besides this, the episode illustrates the context-bound character of conceptual learning processes. The concepts involved, like 'combinations', 'number of all possible cases', etc., are not detached from the application context by means of formal designations; they remain within the framework of the conditions determined by the context without losing their conceptual essence. The fact that 1 is not possible as the sum of pips need not be formalized; however, this could also be called an 'impossible event'. Nevertheless, it would be erroneous to expect the students to derive the conceptually abstract idea of the 'impossible event' as an ideal element within the 'system of all the events' from this directly.

Implications for task systems

The teaching experience with the exemplary task presented here is a typical case in point for how processes of understanding concepts can develop according to the principle of interrelated work between simple models and interpretative contexts in stochastics instruction. The conceptual considerations with regard to task systems for stochastics are summarized in the shape of practical rules:

1) Task systems in stochastics should contain concrete experiments and simulations. The activities may be addressed to individual pupils or to groups of pupils. The task system is connected together by a common formulation of a problem; the individual tasks may require different capabilities from the pupils. A careful organization and coordination of experiments and of simulations or polls helps to orientate these individual tasks according to common aims of teaching.
2) Task systems in stochastics should contain, where possible, a situation requiring a decision. The compulsion to decide, to be subjectively involved in a situation, is an important motivator for pupils.
3) In stochastic task systems, diverse random number generators should be used and compared as objects: from dice to the Galton board and random number table; from loaded dice to real stochastic situations. The requirement to compare such

random number generators compels the study of common properties in the shape of variations, or the analysis of differences in the shape of fundamental changes of the model.

4) An important educational principle of stochastic task systems is the parallel work both on an empirical/experimental and on a theoretical modelling level. Both levels should be related to each other; undue reliance on one of these two modes of work should be avoided. In this sense, the two sides of probability, that is Laplace's probability and probability as relative frequency, should be related to each other as analogous forms of the same concept without, however, being identified.

The analysis of the concept of task system as a potentially important component of teachers' professional knowledge is an example for those approaches in class which, in our opinion, are effective and useful for teaching stochastics. For mathematical knowledge and its acquisition in school a certain complexity of tasks is necessary, and it is only this feature which offers opportunities of gaining access to theoretical knowledge and of shaping the development of theoretical concepts didactically. Other forms which could organize the necessary learning process in class enhancing the theoretical character of probability are small projects, teaching stories, or even historical texts (Jahnke and Seeger, 1986, p.82). In an exemplary way, this is meant to show that curricular materials for stochastics instruction should not and cannot provide 'ready-made' suggestions for the daily practice of teaching. Rather, teaching approaches have to provide orientations and principles and materials for school stochastics which are consistent with the background of the epistemological difficulties of stochastic knowledge. In this sense, task systems are exemplary approaches to stochastics; their principles of construction reflect the relationships between professional teacher knowledge and the curricular elements of stochastics, between classroom activities and the problems of mathematical knowledge.

Bibliography

Andelfinger, B.: 1983, 'Umgehen mit dem Zufall - Ein Erfahrungsbericht aus dem Unterricht (Klasse 7, Gymnasium)', *Stochastik in der Schule* 3 (3), 19-24.

Bentz, H.-J. and G. Palm: 1980, 'Wahrscheinlichkeitstheorie ohne Mengenlehre (Common sense probability)', *mathematica didactica* 3, 167-183.

Biehler, R.: 1985, 'Graphische Darstellungen', *mathematica didactica* 8, 57-81.

Biehler, R. and H. Steinbring: 1982, 'Bernoullis Theorem: Eine "Erklärung" für das empirische Gesetz der großen Zahlen?', in H.-G. Steiner (ed.), *Mathematik - Philosophie - Bildung,* IDM-Reihe Untersuchungen zum Mathematikunterricht, Aulis, Köln, 296-334.

Bromme, R. and K. Juhl: 1985, 'Das Verstehen von Aufgaben aus der Sicht von Mathematiklehrern', *Zeitschrift für Empirische Pädagogik und Pädagogische Psychologie* 9, 1-14.

Brönstrup, H., G. v. Harten, U. Harth, W. v. Lück, K. Osterloff, U. Ries, H. Steinbring, J. Thomas, and P. Weinberg: 1986, *Stochastik in der Klassenstufe 7/8. Einführung in die Elemente der beschreibenden Statistik,* Handreichung für die Gesamtschule, Heft 20, LSW, Soest.

Brönstrup, H., U. Harth, K. Osterloff, U. Ries, H. Steinbring, J. Thomas, and P. Weinberg: 1986, *Stochastik in der Klassenstufe 9/10. Anwendung stochastischer Grundbegriffe am Beispiel des Bernoulli-Experiments,* Handreichung für die Gesamtschule, Heft 21, LSW, Soest.

Cockcroft, W.H.: 1982, *Mathematics Counts*, Her Majesty's Stationary Office, London.

Dinges, H.: 1981, 'Zum Wahrscheinlichkeitsbegriff für die Schule', in W. Dörfler and R. Fischer (eds.), *Stochastik im Schulunterricht,* Teubner, Stuttgart, 49-61.

Dörfler, W. and R.R. McLone: 1986, 'Mathematics as a School Subject', in B. Christiansen, A.G. Howson, and M. Otte (eds.), *Perspectives on Mathematics Education,* Reidel, Dordrecht, 49-97.

Freudenthal, H.: 1975, *Wahrscheinlichkeit und Statistik* 3rd ed., Oldenbourg, München.

Freudenthal, H.: 1983, *Didactical Phenomenology of Mathematical Structures,* Reidel, Dordrecht.

v. Harten, G.: 1985, 'Research and Teacher Training: Co-operative Development of Teaching Materials for Stochastics in Grades 5/6', in B. Christiansen, E.C. Wittmann, M.G. Kantowski, P. Verstappen, L. Burton, G. v. Harten, and G. Brousseau, *Systematic Co-operation between Theory and Practice in Mathematics Education,* Royal Danish School of Education, Copenhagen, 56-64.

v. Harten, G. and H. Steinbring: 1983, 'Randomness and Stochastic Independence. On the Relationship between Intuitive Notion and Mathematical Definition', in R.W. Scholz, (ed.), *Decision Making under Uncertainty,* Advances in Psychology 16, Reidel, Dordrecht, 363-373.

v. Harten, G. and H. Steinbring: 1984, *Stochastik in der Sekundarstufe I,* Untersuchungen zum Mathematikunterricht 8, Aulis, Köln.

v. Harten, G. and H. Steinbring: 1985, 'Aufgabensysteme im Stochastikunterricht. Didaktische Konzepte zur Materialentwicklung und Lehrerfortbildung im LEDIS--Projekt', *Occasional Paper* 71, IDM, Bielefeld.

v. Harten, G. and H. Steinbring: 1986, *Stochastik in der Klassenstufe 5/6. Einführung in die Grundlagen der Wahrscheinlichkeitsrechnung*, Handreichung für die Gesamtschule, Heft 19, LSW, Soest.

Jahnke, H.N.: 1978, *Zum Verhältnis von Wissensentwicklung und Begründung in der Mathematik - Beweisen als didaktisches Problem*, Materialien und Studien 10, IDM, Bielefeld.

Jahnke, H.N. and F. Seeger: 1986, 'Proportionalität', in G. v. Harten, H.N. Jahnke, T. Mormann, M. Otte, F. Seeger, H. Steinbring, and H. Stellmacher et al. (eds.), *Funktionsbegriff und funktionales Denken*, Aulis, Köln, 35-83.

Loève, M.: 1978, 'Calcul des Probabilités', in J. Dieudonné (ed.), *Abrégé d'Histoire des Mathématiques 1700-1900*, vol. II, Hermann, Paris, 277-313.

Otte, M. and V. Reiß: 1979, 'The Education and Professional Life of Mathematics Teachers', in B. Christiansen and H.-G. Steiner (eds.), *New Trends in Mathematics Teaching*, UNESCO, Paris, 107-133.

Pólya, G.: 1967, *Schule des Denkens*, Francke, Bern.

Steinbring, H.: 1980, *Zur Entwicklung des Wahrscheinlichkeitsbegriffs - Das Anwendungsproblem in der Wahrscheinlichkeitstheorie aus didaktischer Sicht*, Materialien und Studien 18, IDM, Bielefeld.

Steinbring, H.: 1984, 'Mathematical Concepts in Didactical Situations as Complex Systems: The case of Probability', in H.-G. Steiner and N. Balacheff (eds.), *Theory of Mathematics Education (TME) ICME 5 - Topic Area and Miniconference, Occasional Paper* **54**, IDM, Bielefeld, 56-88.

Steinbring, H.: 1985, 'Zur Behandlung stochastischer Grundbegriffe im Unterricht - Eine vergleichende Falluntersuchung zwischen Grund- und Erweiterungskurs einer 7. Klasse', in *Beiträge zum Mathematikunterricht*, 302-304.

Steinbring, H.: 1985, 'Dimensionen des Verhältnisbegriffs im Stochastikunterricht', *mathematica didactica* **8**, 217-238.

Steinbring, H.: 1989, 'The Interaction between Teaching Practice and Theoretical Conceptions - A Cooperative Model of In-Service Training in Stochastics for Mathematics Teachers (Grades 5-10)', in R. Morris (ed.), *Studies in Mathematics Education. The Teaching of Statistics*, vol. 7, UNESCO, Paris, 202-214 .

Winter, H.: 1983, 'Zur Problematik des Beweisbedürfnisses', *Journal für Mathematikdidaktik* **1**, 59-95.

Heinz Steinbring
Institut für Didaktik der Mathematik
Universität Bielefeld
Postfach 8640
DW-4800 Bielefeld
FR. Germany

Chapter 6

Computers in Probability Education

R. Biehler

This chapter is concerned with the impact of computers on probability in general secondary education. Mathematics educators have been producing ideas for using computers and calculators in probability education for two decades. Although there are many teaching suggestions, empirical research on this topic is uncommon and critical reports of practical experience rarely go beyond an enthusiastic description. A critical review of ideas, software and experience which would be helpful for further research and development is the major objective of this chapter. We will deal with pedagogical aspects, the subject matter and its change, and the role of changing technology. Various approaches will be reviewed; computers used as general mathematical utilities, simulation as a scientific method, and simulation for providing an empirical background for probability. Graphical methods may enhance the idea of visualization. The emphasis is on general orientation in the field.

1. Computers and Current Practice in Probability Teaching

If we asked informed experts in educational computing 'What is really going on with computers in the probability classroom?', most of them would say 'Not very much at the moment.' The inherent pedagogical problems of a computer orientated approach are dealt with below.

Pedagogical problems and perspectives

The tardy integration of computers is also true for mathematics teaching in general. Even in England, where there is good educational software, the use of the computer in probability is still uncommon; in many other countries, available software is inferior. Occasionally teachers develop short and somewhat unprofessional programs, or pro-

R. Kapadia and M. Borovcnik (eds.), Chance Encounters: Probability in Education, 169–211.

gramming is done in the classroom together with the students. Apart from the lack of software, many other familiar factors are responsible for this situation.

In many countries, probability has entered the curriculum only recently. Teaching probability is a difficult task. It may seem counterproductive to multiply difficulties by introducing the computer as a further element which needs organizational effort and additional competence. It may seem strange that educators consider new potential content in probability which is now accessible with the support of computers, even though the very elementary ideas of probability are neither well understood nor well taught. On the other hand, the new technology might help to overcome central learning difficulties. Two aspects of probability education yield a promising starting point for further research into targets of difficulty which appropriate pedagogical use of computers may help solve:

(i) 'lack of experience'. Time constraints and limited resources do not allow for sufficient opportunities for gaining experience to support the learning of probability. Learning environments which allow computer simulation as well as exploration of real data may be of considerable help here.
(ii) 'concept-tool gap'. There is a gap between the intended generality of the probability concept and the system of operations and tools students actually use. Solving problems by simulation or substituting analytical methods by computer-based graphical and numerical methods makes a new range of problems and realistic situations accessible to students' activities.

The two targets of difficulty are related to two complementary approaches to the use of computers, namely as idealized objects for exploration and as cultural tools for working on problems. The new technologies are perhaps more relevant for probability than for other parts of the mathematics curriculum which have fairly unified methods. As probability is an applied subject, the objects referred to (i.e. random situations and processes, decision situations) and their comprehensive study have an importance of their own; this calls for a greater diversity of methods and tools. Moreover, there are numerous problems which are conceptually simple but very time-consuming; just think of collecting the results of a sample of 20 experiments with 100 throws of a die.

Using computers in teaching will also contribute to an understanding of the new technologies and of their competent use. For instance, it is not adequate to consider simulation merely as a method of teaching; rather, the use of simulation outside school is also considered relevant if one takes a broader, cultural view of educational innovations.

What are the challenges and perspectives for mathematics education in general, and for probability education in particular? Pollak (1989) identifies problem areas where secondary mathematics education should contribute to developing 'intelligent citizenship':

* uncertain situations, in particular low-probability and high risk events
* reasoning from data
* intelligent planning and optimization
* systems thinking (appreciation of complexity, unforeseen consequences, instabilities)
* thinking discretely or algorithmically
* understanding the possibility of modelling situations in a structural, quantitative and analytic way

Most of these areas are related to probability and to the use of computers. Sometimes computers cause the problem, e.g. an abundance of data; or they are involved in analyzing the problem, e.g. stochastic models for the spread of *AIDS* or risk analysis of nuclear power plants. In other words, people are confronted not only with 'primary' randomness but with culturally mediated randomness, i.e. with other people's attempts to control, understand, and use uncertainty with available tools.

As children have experience with randomness in games on home computers, one cannot postpone the treatment of computer-generated random numbers beyond school level because of their mathematical complexity. These situations, therefore, call for developing a new conception of the subject matter of probability taking the new technologies into account. The use of computers outside school and related changes in social practice and academic disciplines provide relevant background knowledge for curriculum designers and teachers alike. Some important developments will be briefly described now.

Changes in probability, statistics, and in their applications

Computer technology has enriched the tools available to handle random situations. Much professional software has been produced; among them are statistical packages like *SPSS*, interactive statistical systems for data analysis like *S*, specific simulation languages like *Simula*, and modelling systems like *Stella*. They intend to support different users, working styles, and objectives in various areas of science and practice. Their educational potential is as yet largely unexplored.

New tools are related to new ways of thinking about random situations. This applies to individual humans or to whole scientific communities. Two domains of rapid development are data analysis and simulation, which, at the same time, are key notions in many disciplines. The accelerating quantification of many parts of society is greatly influenced by the new technologies. The use of computers has partly revolutionized the practice and theory of analyzing data, particularly through Exploratory Data Analysis (Biehler, 1985). Simulation, although perhaps less present in everyday life than data has enriched the methodology in many scientific applications. The following categories from Howson and Kahane (1986, pp.2-12) are useful to describe the influence of information technology on probability.

New and revived areas of research in statistics and probability.

* simulation and modelling
* random number generators
* stochastic processes
* probabilistic analysis of algorithms
* exploratory and graphical data analysis
* multivariate data analysis and statistics
* computer-intensive methods (like bootstrap methods)
* expert systems for decision making under uncertainty

Experimentation and proof in mathematics. Computers support an interactive and experimental style of working with models and data. This constitutes a new type of scientific method which still has to find its role with regard to more traditional strategies. Computers can be used for supporting conjectures by numerical or graphical evidence; for supplying counter examples; and for assisting the process of proving by calculation and enumeration.

In such a context, proofs can be more clearly considered with regard to their role for explanation and insight. More controversial are aspects such as probabilistic proofs, where it can be shown, for instance, that a number is prime with a very high probability. Are such methods acceptable within mathematics? This problem was partly responsible for the slow acceptance of the Monte Carlo method by statisticians in the early days; compared to pure mathematics, they were considered second class. Tukey (1962) attacked this mathematical purism fiercely. The opinion that analytical methods are superior for their greater generality and certainty has now shifted; simulation is no longer a last resort but is a method of its own right. These scientific developments broadened the understanding of mathematics and may help to overcome a narrow view of school mathematics. However, this liberal attitude is not yet common among mathematics teachers.

Iterative methods. With the advent of computers, iterative methods have acquired a prominent place in many domains of mathematics and its applications. Simulation in the sense of the Monte Carlo method is a wonderful example for the power of iterative methods. It may be, in essence, described as using pseudo-random numbers to simulate an experiment in order to arrive at a quasi empirical estimation of some unknown probability. Ulam and v. Neumann pioneered this approach in the fifties, although there had been isolated uses of such techniques before. Numerous calculations are necessary to apply the method; so it is greatly facilitated by computers which have been essential for its growing success. Since then, so-called computer-intensive methods in probability and statistics including iteration and simulation have experienced a very rapid development (Efron, 1979).

Algorithms. Algorithms as a part of mathematics are penetrating probability on a concrete and on a meta-level; there are algorithms for generating random numbers, for standard tests, and for probability distributions; more significantly, instead of formulae for calculating the coefficients of least squares, we now find algorithms for an iterative or recursive calculation of robust lines or for smoothing curves. In other words, algorithms are independent means of understanding and representing mathematical knowledge; they cannot simply be substituted by formulae and theorems.

Simulation and symbolic modelling. Random situations can be represented by mathematical models, by physical models (like urns) and by computer programs. 'Computational' models have a dual nature: they are symbolic models like mathematical models (represented in a formal language) and, at the same time, they are physical models (represented in a real machine). Computer programs permit the extension of probability modelling to new domains via more complex and realistic models. Options such as easily changing the assumptions of the model, making further experiments, changing the way generated data are analyzed etc. are only available in computer environments.

In this respect, computational modelling is closer to the hypothetical and exploratory features of traditional mathematical modelling than physical modelling. Moreover, computational models are often robust against a change of assumptions. A change of one or two program lines provides a sufficient adaptation. Contrary to that, analytical models are based on specific assumptions like the normal distribution and it is difficult to analyze the effect of slight deviations from them. Simulation has also become a research tool even in theoretical statistics, new disciplines like computational probability have emerged (Freiberger and Grenander, 1971).

Visualization, graphical methods and other representations. We can speak of a scientific revolution with regard to the use of graphs in statistical analysis; dramatic changes are also visible with regard to simulation. While in the early days of the Monte Carlo method the ability to rapidly execute iterations was exploited to get a list of numbers, modern technology makes it possible to represent the model and its related system on different levels of abstractness on the screen, each of which may serve as an independent source of knowledge to explore the model. Examples are graphical displays of random walks in the plane, pictures of cellular automata and graphical displays of numerical data which have been generated during a simulation. The use of computer graphics is related to the use of classical media and tools like static print graphics, video and film. Computers add a specific type of adaptability and interactivity.

Changing technology and its influence on pedagogical ideas

Changing technology has had a major influence. Initially, calculators were used for numerical work such as evaluating formulae of combinatorics; simple simulations became possible with programmable calculators. Programming languages marked a big step beyond the idiosyncracy of machine languages; for a long time *BASIC* dominated probability education whereas other languages remained largely unexplored. The suggestions differed with regard to the students' participation in writing, modifying or reading the *BASIC* code and, consequently, with regard to the pedagogical opportunities and difficulties.

A significant new step is the micro with a graphic screen which opens up new visual facilities, such as dynamical displays of stochastic processes on several levels of abstractness, e.g., from showing balls falling through a Galton Board on a screen to plotting relative frequencies while simulating the tossing of a coin. In addition, an extensive interaction with static displays becomes a possibility, e.g., varying parameters of binomial distributions as a means of studying this distribution family.

Another step forward is marked by more sophisticated educational software and consciously designed learning environments with a polished user interface. A review of interesting items will be integrated in this chapter. Further opportunities are offered by general mathematical tools such as spreadsheets, supercalculators, modelling systems and systems for statistical data analysis. The latter can support extensive use of real and simulated data in elementary probability.

How can the relation between pedagogical and technological progress be assessed? We can already observe some difficulties at the software design level. For instance, some educational software shows pictures of random needles flying across the screen to help in visualizing Buffon's needle problem. Observing such animation is quite a different demand as compared to programming models on a simple calculator; instead of constructing scientific models, students watch a movie. There is a need for design principles which safeguard against the temptation of the advanced technology taking over those processes which are considered important for the student's intellectual development. In the following, the dependence on technology as well as a critical comparison with the use of other tools and non-computer approaches is one of the basic perspectives.

2. Computers as Mathematical Utilities

There are many modes of using computers which are not specific to probability, namely when formulae, equations, functions, numerical calculation and sequences are used together with software. Traditional tools like books of tabulated functions (like the binomial distribution) as well as the 'closed formula' can be supplemented by using algorithms to represent and explore relationships. The 'birthday problem', Bayes' formula and the binomial distribution will be used to illustrate some of the new possibilities.

The birthday problem

As an example, we explore the birthday problem: if 30 people meet in a room, what is the probability p that at least two of them have their birthday on the same day of the year? The solution is p=0.71 if we make the assumption of a uniform distribution of birthdays. In common teaching approaches, the emphasis is on solving the combinatorial problem by computing p, and developing an intuitive understanding of why p is so high with only 30 people. In a computer environment, many other aspects can be discussed.

A general formula for the required coincidence probability p(s) is

$$\mathrm{p}(s) = 1 - \frac{365 \cdot 364 \cdot \ldots \cdot (365 - s + 1)}{365^s}$$

A simple *BASIC*-program for computing p(s) iteratively is

```
10      INPUT N,S : Q = 1
20      FOR I = 0 TO S - 1
30      Q = Q*(1 - I/N)
40      NEXT I
50      PRINT 1 - Q
```

For n=365, p(s) is computed. With the help of this program many interesting questions can be explored, such as

- for which number s of persons is the coincidence probability p(s) larger than 0.5?
- the functional dependence p(s) on s which is not proportional,
- the analogous 'birth-month' problem

The example illustrates how short programs can help to bridge the 'concept-tool-gap'. Generalizations and variations of the original problem can be studied easily using the symbolic representation in the program, especially those which are tedious to study with formulae alone.

Let us look at the birthday problem from another angle; instead of a formula for the relation between s and p(s), another alternative is to aim directly and conceptually at an iterative solution. Let q(i) be the probability of no common birthdays in a group of i people and p(i) = 1 - q(i) the probability of at least one common birthday. For i=1, q(1)=1. Imagine we have already a group of i people ($i < n$, n=365) without a common birthday and another person enters the room. The probability that his/her birthday equals one of the i persons is i/n. Thus we find (see also fig.1)

$$q(i+1) = (1 - \tfrac{i}{n}) \cdot q(i)$$

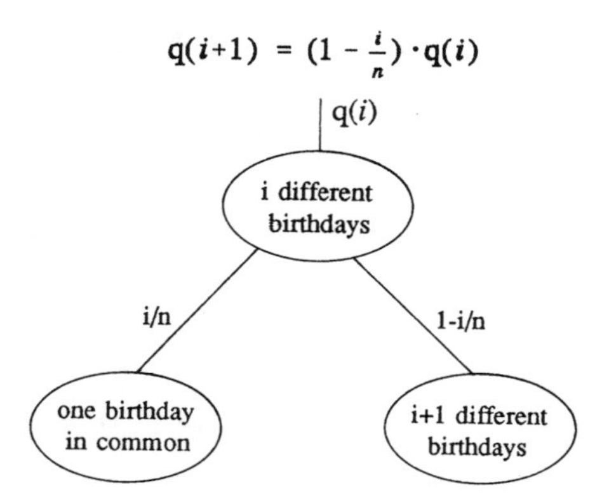

Fig.1: Tree diagram for the next person to enter

Instead of using *BASIC*, this iterative formula can be put onto a spreadsheet with its facilities for generating tables and graphs, and where iteration can be performed by copying formulae. The data in table 1 were generated in two steps. The formula above was inserted for i=2. Then this formula was 'copied' to all the other rows. Columns could be plotted against each other to study the functional relationship. If we change n=365 to n=12 or 52, everything is automatically recalculated and we have solutions for the birth-month or the birth-week problem. Another advantage of the spreadsheet over the above *BASIC*-program is that all intermediate results are shown. This iteration is now no longer only a computational formula but has a meaning in the situation. The program is not a simple computational tool; rather, the concept of iteration influences the thinking process from the beginning. Such algorithmic or procedural thinking is often promoted because of its new cultural value or because it can attract students who have difficulties with other styles of mathematical thinking. In other cases, the algorithmic representation may substitute for a formula which may not even exist.

i	q(i)	p(i)	i	q(i)	p(i)	i	q(i)	p(i)
1	1	0	18	0.65	0.35	35	0.19	0.81
2	1.00	0.00	19	0.62	0.38	36	0.17	0.83
3	0.99	0.01	20	0.59	0.41	37	0.15	0.85
4	0.98	0.02	21	0.56	0.44	38	0.14	0.86
5	0.97	0.03	22	0.52	0.48	39	0.12	0.88
6	0.96	0.04	23	0.49	0.51	40	0.11	0.89
7	0.94	0.06	24	0.46	0.54	41	0.10	0.90
8	0.93	0.07	25	0.43	0.57	42	0.09	0.91
9	0.91	0.09	26	0.40	0.60	43	0.08	0.92
10	0.88	0.12	27	0.37	0.63	44	0.07	0.93
11	0.86	0.14	28	0.35	0.65	45	0.06	0.94
12	0.83	0.17	29	0.32	0.68	46	0.05	0.95
13	0.81	0.19	30	0.29	0.71	47	0.05	0.95
14	0.78	0.22	31	0.27	0.73	48	0.04	0.96
15	0.75	0.25	32	0.25	0.75	49	0.03	0.97
16	0.72	0.28	33	0.23	0.77	50	0.03	0.97
17	0.68	0.32	34	0.20	0.80	51	0.03	0.97

Table 1: Probabilities of coincidences in the birthday problem

A more radical top-down approach is to go directly from the problem to the formula using a black box program. Having studied the relation between s and p(s), the question of how to derive p(s) might be addressed. Such a reliance on other people's algorithms as a 'primitive' conflicts with traditional mathematics education. The traditional core of probability is to learn and execute the techniques and algorithms which have been developed to compute unknown probabilities from known ones. However, maintaining this bottom-up attitude as a major principle of secondary mathematics education is severely challenged by the computer environments outside school where the reliance on algorithms which are not transparent in detail, is the norm.

The above calculations are based on the (tacit) assumption of a uniform distribution of birthdays over the year. When students are asked to construct a simulation model, this assumption has to be dealt with explicitly. The computer can be used to study various other models and compare such models with real data; a simulation program for the uniform distribution can easily be modified to any kind of distribution. The assumption of uniform distribution may be checked by investigating real data. If we have the number of births $n(i)$ for each day no. i of a year and n as the total number of persons, plotting the residuals $\frac{n(i)}{n} - \frac{1}{365}$ against i gives information about the deviations from the assumption. In an adequate computer environment, a shift of interest to assumptions of the model, their change and comparison with real data would be easily possible.

Exploring Bayes' formula

Bayes' formula is another illustrative example where the computer can help with the 'concept-tool gap'. In a sense, this also holds for professional statistics. The availability of computers was one of the factors which supported new interest in Bayesian methods, because they are computationally intensive. Computers can now be fruitfully used for supporting subjectivist aspects of probability in teaching. Programs can be used to enrich the learning, by exploring the formula with varying prior probabilities and by iterative application of Bayes' formula to sequences of real data.

Bayes' formula is often underestimated as a mere formula for computing unknown probabilities from known ones. It can play a more important role in probability education as a tool for making probability judgements; psychological research gives additional support for this (see Chapter 7).

A popular topic is the use of Bayes' formula with medical tests. For illustrative purposes we use a recent discussion in the magazine *CHANCE* 1 (1988, p.9) on the disadvantages of mandatory *AIDS* tests. Let us define the events

$V+$: a person has the virus, $V-$: a person does not have the virus;

$T+$: test was positive, $T-$: test was negative.

Suppose that the test is very reliable so that for an individual with the virus, the probability of a positive outcome is very high, $P(T+|V+)=0.999$; and likewise, for an individual without the virus, the probability of a negative outcome is also high, say $P(T-|V-)=0.99$. Using an estimate of 1.5 million Americans with the virus, we have

$P(V+)=0.006$, approximately. Then, Bayes' formula yields the probability that someone with a positive test result actually has the virus

$$P(V+|T+) = \frac{P(T+|V+)\cdot P(V+)}{P(T+|V+)\cdot P(V+) + P(T+|V-)\cdot P(V-)} = 0.376$$

Thus, under mandatory testing of the general population with the assumed probabilities, almost two-thirds of those with a positive test would actually not have the virus. The large difference between $P(V+|T+)$ and $P(T+|V+)$ is surprising for many people, who cannot make the conceptual distinction between both conditional probabilities.

Now, a programmable calculator or computer is used to do a sensitivity analysis ('what if'-scenario), i.e. how does the result depend on the assumptions? What could be done to increase $P(V+|T+)$? How can we explain the unexpectedly low probability of 0.376 for having the virus given the test is positive? Such an exploration could be more easily done in a spreadsheet environment or in a system like *MathCAD*. In the latter, the formulae and assignments of the numerical values can be typed on the screen and whenever something is altered in one place of the screen, say the estimation of $P(V)$, all the other values are recalculated. A graph of dependencies on input parameters might reveal insight. The new possibility of sensitivity analyses is an important aspect of applied mathematics, greatly facilitated by computers.

The iterative application of Bayes' formula to sequences of real data is the core of Bayesian inference. An obvious use consists in just showing how a 'Bayesian machine' works on sequences of real or simulated data, for instance through displaying or plotting the table of changing probabilities. For instance, Riemer (1985) developed an approach which is supported by material and classroom trials and intends to restructure the lower secondary probability curriculum aiming at an earlier introduction of basic statistical ideas. From the beginning, the concept of a hypothesis is central and students are asked to formulate different hypotheses for partly symmetrical dice and other experiments; experiments are evaluated according to Bayes' formula.

Binomial probabilities

The formula for binomial probabilities represents several functional relationships which are relevant for probability:

$$P(X=k) = \binom{n}{k}\cdot p^k\cdot(1-p)^{n-k}$$

- the distribution for fixed p and n (dependence on k);
- the effect of sample size n and the probability p;

- the effect of varying p on the probability of $X=k$.

Binomial coefficients hide a basically recursive definition which does not fit into current curricula with their preference for functions of one variable defined by an algebraic expression. This reduction hinders a more general understanding of functional thinking (v. Harten et al., 1986).

In schools, several tools and visualizations are already used to improve understanding: the recursive relation can be studied as numerical examples of Pascal's triangle before the formula is introduced; tables for binomial probabilities and numerical approximations are used; the functional dependencies on k, p and n are visualized by graphs in textbooks. Computers can use this as a starting point.

For instance, the software package *Advanced Level Studies: Statistics* provides programs to make such activities more efficient; see Holmes (1985) for a description of the software. The possibility of graphing the distributions is powerful; the comparison of binomial probabilities to real data, otherwise tedious, becomes very easy. It is also possible to select data from a library which can be extended by teachers and students. So the programs support the comparison of the binomial distribution to real data which is often neglected in non-computer environments.

Computer implementation may support a more active numerical and graphical exploration of formulae; graphs can become exploratory tools for problem solving which is hardly possible without computer support, but this potential is too rarely exploited by current educational software.

Concealing the computational algorithms in such menu driven programs relies on students accepting that calculation is done according to formulae without knowing them. This is similar to relying on calculators which compute standard formulae by a single key stroke. This possibility has often been criticized, for instance when people use statistical packages without any understanding. Under certain conditions, however, an educational opportunity can help to handle the 'concept-tool gap'. It may be possible that a metaphorical understanding of a 'black box' can guide a reasonable application.

For instance, the distribution of the number of successes in n Bernoulli trials could, in principle, be calculated by generating all the cases (the whole sample space) and counting the number of successes in each case to obtain the distribution. This algorithm could be accepted as a general solution to the problem, and students can use a binomial probability program with such a metaphorical understanding, just as people outside school use programs for calculating the t- or the F-distribution with an understanding of the statistical meaning of the distributions, but without remembering or knowing the formulae.

Programming languages and other tools

The question of which general mathematical tools can be used with success in the probability classroom has to be raised. Several types of such tools are available to support learning and applying probability in principle, namely programming languages, spreadsheets, general mathematical systems like *MathCAD*, *Derive*, or *mathematica*, modelling and simulation systems, or systems for statistical data analysis.

A choice depends on several criteria:

(1) availability in the school

(2) usefulness for applications outside probability

(3) adaptability for probability problems

(4) ease of use for the learners

A major problem with regard to mathematical tools is that each one requires specific knowledge, for both students and teachers. As time in school is very limited, more complicated tools for probability education cannot be chosen without taking into account which tools should be used in the entire mathematics curriculum. This is why the usefulness outside probability is important. The relative merits of the different program types have still to be explored. Programming languages, especially *BASIC*, were explored in education as they are easily available, easy to use, and frequently used outside school. Therefore, *BASIC* is acceptable for many educators. Restructuring of school probability from an algorithmic point of view has been an additional argument for programming languages. However, experience of the last 15 years have shown that programming as a student skill is difficult to achieve and may shift the focus away from those conceptual issues which are considered essential for probability.

Nevertheless programming languages have their advantages. They are not mere technical devices for encoding algorithms, but tools for thinking on the one hand and mediators of computer models on the other. For instance, many languages support the model of a machine which can iteratively execute commands and at the same time provide notions for formulating iterative processes. In contrast, menu driven programs have a preselected range of options which facilitate the orientation of the users but they lack flexibility of choices and sequels. Programming languages require a lot of special commands; in return they offer a highly flexible tool. A short program say on computing binomial probabilities may be used as a module which is adaptable to many further problem solving activities.

The pragmatically established predominance of *BASIC* has prevented a more intensive study of other programming languages with regard to probability in education. For instance, when using a general purpose programming language with the facility of defining procedures, a library of program modules might gradually expand as a result of class work (Klingen, 1977).

Matrix or vector oriented languages are even better fitted to the needs of applications in statistics and probability; if the commands are directly executable, interactive work is enabled. Among computational statisticians outside school, the programming language *APL* is often discussed. With regard to schools, Alvord (1985) developed and evaluated a course in elementary ideas in combinatorics using *APL*. The strange notational system of *APL*, which is different from algebraic notation, has been criticized. Newer matrix oriented languages like *PC-ISP* or *S* have been a reaction to that critique; they also enable a style of work best called 'improvisation'.

Spreadsheet environments with its tabular representation necessitate a different style of thinking but enable easy access to graphics, too. Their availability and use in other subjects is a clear advantage, although their handling of data vectors is less convenient for probability than in the vector oriented languages.

Still another type of utility software combines the advantages of a library of statistical methods, interactive graphics, options for generating random numbers and easy access to calculation, namely easy to use tools for statistical data analysis. *StatView* and *Data Desk* are two good examples, running on an *Apple MacIntosh* environment. For a detailed survey of such tools see Biehler and Rach (1990).

An essential question is whether such tools can be adapted and used for purposes in school or whether it would considerably help if it were possible to develop software tools especially for mathematics education in schools, including probability education. Some points of view on this problem can be found in Biehler (1991) or Kaput (1988).

3. Simulation as a Problem Solving Method

The use of simulation in probability education is often suggested. An abundance of examples is already published. But the different roles, goals and pedagogical perspectives for simulation have not yet been clearly analyzed and distinguished. A basic distinction is between the use of simulation as a method for solving problems, similar to the professional use outside school, and the use of simulation to provide model environments to explore, which compensate for the 'lack of experience'. The latter aspect will be discussed in the next section, here we analyze different uses of simulation for problem solving, reaching from a simple numerical method for approximately determining unknown probabilities to an integrated part of modelling. Simulation can bridge the 'concept-tool gap' in secondary education, by serving as a simpler replacement for difficult or non existent analytical methods.

Integrating simulation, algorithmics and programming

Engel's approach to integrate simulation, algorithmics and programming is still influential. In searching for interesting applications of programming in mathematics he concluded (1975, p.95)

> "For me it was a pleasant surprise to discover that probability is even more suitable than number theory... At the high school level, computers are ... misused as super-slide-rules. But a computer is really a simulator of processes, ... Simulation programs are the most instructive programs. The computer is instructed to imitate some process, for instance, to play 1000 crap games."

He has developed a number of interesting examples, connecting the algorithmic approach with embedding simulation in the context of computational or symbolic modelling (Engel, 1977; 1985).

His basic metaphor for relating programming to simulation is *constructing probability machines* in the computer. He suggests starting from the *RND*-command as a primitive and using a programming language as a *construction set* for defining more complex machines. The algorithmic representation of the process is further enhanced by relating the various *RND*-commands to corresponding spinners as their iconic counterparts. These spinners are used both as metaphors to explain what the *RND*-command does and as an intermediate step in the process of symbolic modelling.

Another dimension in Engel's approach is his emphasis on simulating *processes*. This is a conceptual framework for interpreting random experiments which is ignored in conventional approaches. In the framework of processes, generalizations and variations are possible which are much more difficult with (static) probability spaces. For instance, the rules of crap involve waiting for certain events and planning further playing according to the result. This can be quite easily expressed in the primitives of an algorithmic programming language - no theory of stochastic processes is necessary for that.

Engel uses a complementary representation for the structure of a process. It is possible to visualize a process as a random walk on a finite graph. This is illustrated by the following problem (Engel, 1977, pp.163). Abel says to Kain: We want to toss a coin with a side zero and a side one until one of the sequences 1111 or 0011 appears. In the first case, you win, and in the second, I win. The game is fair, because both sequences have the probability of 1/16. Is Abel right? What are his winning probabilities?

Instead of analyzing the probability spaces, the underlying process is studied. This is done by algorithmic thinking and not by global relations of Markov processes. A possible algorithm is:

(1) Toss a coin 4 times

(2) Note result of the last 4 tosses as x

(3) If x=1111 : Kain wins

(4) If x=0011 : Abel wins

(5) Else : Toss a coin once again, go to (2)

This algorithm can be coded and implemented. A simulation shows that Abel's winning probability is much higher than 0.5 ! As usual, a simulation provides a number and points to a fact which has to be explained by other means. The graph in fig.2 can be used for explaining why the game is not fair.

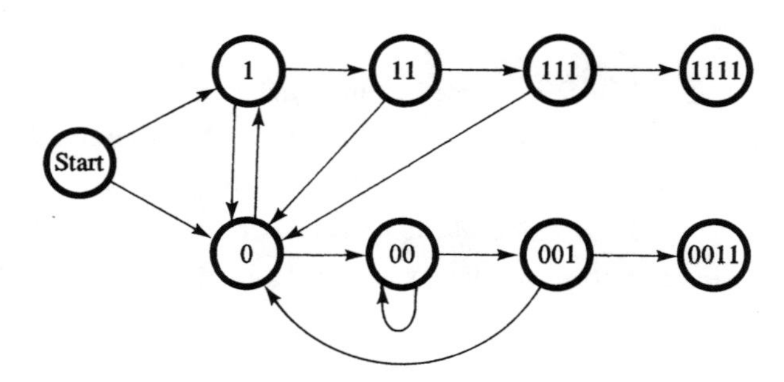

Fig.2: States and transitions in 'Abel Kain'-problem

The circles represent the different possible states of the process, which can end either in 0011 or in 1111. Each arrow in the diagram represents the transition probability 1/2. For instance, after the state 11, it is equally likely to get into 111 or into 110. But with regard to the two final states 1111 or 0011, the state 110 is equivalent to 0. Similar reasoning explains the other arrows in the graph from which a qualitative explanation is possible: it is more likely to fall from the upper line to the lower line than vice versa. The analysis necessary for constructing the graph is much more demanding than the programming to get a numerical result although it does not involve a complicated theory.

There is a tension between the algorithmic view and the interpretation of simulation as modelling; the question is whether the goal is to get an algorithm which solves the problem or to get a model which serves for further explorations. The concept of a simulation algorithm can be linked to the process of physical simulation; various classroom suggestions exploit this aspect. But exaggerating the algorithmic standpoint may be misleading. More recent conceptions of modelling distinguish between the following aspects of simulation:

(1) Formulating the model
(2) Simulating the model
(3) Analyzing the results of the simulation
(4) Validating the model

Recent software tools like *Stella* and *DMS* reflect these distinctions and thus support the integration of simulation into an overall conception of modelling.

Simulation as an alternative to solving problems analytically

Discussing the use of non-computer simulation in secondary probability education has a long tradition. A major reason has been that using simulation is easier than working with mathematical models for many students. We analyze some ideas of this different thinking tradition now, in which the computer plays a minor role though it may facilitate the simulation. For analytic reasons, it is helpful to distinguish major points in favour of simulation:

Representational aspect. Students can think and formulate models in concrete terms like urns or spinners; there is no need to express models symbolically in terms of probability spaces.

Computational aspect. Processing the generated data, say for estimating an unknown probability may be easier for students than using analytical/ combinatorial methods; it may be the only way to include more complex or realistic probability problems in the curriculum.

Concept of model aspect. To solve a probability problem (such as the birthday problem) by simulation one has to design an experimental set-up and think of a model first instead of starting with blind calculations.

In the United States, a number of research studies and practical suggestions have been developed along these principles. One of their major specific ideas is the pedagogical comparison of analytic methods with Monte Carlo methods. Some articles in the *NCTM Yearbook* reflect this tradition (Inhelder, 1981; Travers, 1981; Watkins, 1981). The suggestions are partly related to a series of empirical studies in which teaching with the Monte Carlo Method was compared to the usual analytical method (Atkinson, 1975; Holmes, 1968; Sanders, 1971; Shevokas, 1974; Simon et al. 1976). A recent textbook (Travers et.al., 1985) extends this tradition, including the analysis of real data. There is also the recently published simulation language *Resampling Stats* which explicitly refers to this tradition.

The main reason for using Monte Carlo Methods in teaching is seen in its potential to bridge the 'concept-tool gap'. It is often claimed that this quantitative extension is also a qualitative leap: with usual methods, it is hard to get beyond conventional games of chance and simple problems related to this domain. By using Monte Carlo simulation, even without computer support, it is possible to attack more realistic situations and to solve sophisticated problems in a unified way. This may give students a feel of control over probability (Shevokas, 1974, p.163; Watkins, 1981, p.209).

Atkinson (1975) empirically compares the success of a Monte Carlo course with one based on the analytical approach, showing the former to be superior in certain respects. In view of the simple problems involved, this is quite astonishing. However, the analytic course did not use simple tools like tree diagrams at all but drew heavily on set theory. It might be better to compare and combine approaches to the theoretical and experimental aspects of probability rather than setting simulation against analytic methods. We believe that the advantages of analytical methods for *explaining* phenomena may become clearer when used in connection with simulation. It is not surprising that this function of analytical methods is often 'overseen' as in school mathematics everything tends to degenerate to mere calculation. Regarding simulation only as the easier method of calculation as is done, for instance, by Inhelder (1981, p.220), misses essential points. The purism in mathematics rejecting simulation as only an initial step has begun to change. This gives additional support to a general approach that views simulation as a real extension of analytical methods while remaining aware of the specific features and partial advantages of both.

Conventional teaching neglects the relations between a mathematical model and a random situation. The 'American tradition' considers the construction of a model for a simulation as an important intellectual challenge. In general, however, their view of modelling is limited. The student tasks suggested are largely written from the traditional point of view; the focus is on initial formulation of a predetermined implicit model and probability theory is concerned with calculating or estimating unknown numerical values of probabilities. The analysis of real data, questions of model validity, justification of assumptions etc. are ignored in publications and textbooks. (Gnanadesikan et al. 1987; Watkins, 1981).

A more advanced substitution of analytical methods by simulation is apparent in two recent textbooks. Landwehr et al. (1987) introduce confidence intervals on the basis of the theoretical distribution of the proportion in a sample of size n. This theoretical distribution is said to have been derived from very long run simulations; the simulated result is used as a kind of mathematical definition. Travers et al. (1985) use results of simulations as the theoretical counterparts to applying statistical tests on real data.

However, the central idea of *substituting* analytical methods by simulation limits the pedagogical perspective. Using adaptable random devices such as urns or spinners as cognitive models is an important first step in modelling a situation; this can be followed by translating the concrete model into an abstract mathematical formulation, which is then the basis for applying the probability calculus. For instance, there are whole books based upon this approach using urns or spinners (Freudenthal, 1975). An alternative way could be to simulate an urn model on a computer. From this perspective, the use of non-computer simulation is an important step in the students' cognitive development in so far it helps to build up their cognitive repertoire of models which remain relevant for model formulation even when these models are analyzed theoretically or by computer programs.

The potential of computer-aided simulation

The major drawback of simulation without computers is the time needed to perform the physical experiment. One solution is to construct more efficient sampling devices. A 'sampling paddle' is an effective tool for rapidly generating samples of size five, ten or fifty; it is a board with a number of hemispherical hollows which serves to ladle samples from a box containing several thousand wooden balls.

The use of computers supports the tendency to remove other random devices from the mathematical laboratory. Although it is difficult to generalize about the relative merits of computers because much is dependent on the software and the associated pedagogical strategy, the following aspects seem to be central advantages of the computer approach:

(1) the number of repetitions is easily increased so that uncertainty and variation in the results can be reduced; new kinds of patterns become detectable,

(2) an extensive exploration is possible by changing the assumptions of the model, making further experiments, changing the way generated data are analyzed etc.,

(3) new and more flexible representations are available to express models and stochastic processes and display data with graphical facilities.

In this respect, computer simulation or computational modelling is much closer to the hypothetical and exploratory features of traditional mathematical modelling than physical modelling. The use of computer simulation might strengthen the applied nature of probability. However, a reasonable use of computers is necessary, otherwise the advantages of the simulation method may get lost in the computer environment.

From the *computational aspect*, computer simulation seems to be more efficient than non-computer simulation. However, students are now asked to specify which computations should be performed. Dependent on the software, this may be a source of new problems. Moreover, if we take classroom processes and student motivation into account, we may also come to a slightly different judgement. Non-computer simulation can easily be used to introduce more practical and co-operative work in the classroom which has an educational value in itself, even if it may not be the most efficient way of problem-solving.

A major problem is related to the representational aspect: How is the computer used for expressing and representing probability models? This problem was already discussed in the early days of computers in probability education. Although a simple simulation language with 20 commands which was designed for solving the standard textbook problems in probability was available (Simon and Weidenfeld, 1973), Shevokas (1974) suggests the use of programs as a black box where the students do not know anything about the details of the program, but have a general awareness of its capability, for example, that it generates data like a fair die. The user interface is very simple; only a few parameters specifying a model are allowed as input. Her argument is fairly clear: if simulation helps to avoid algebraic notation, a formal notation through formal languages is counter productive.

However, even with these programs, the group of students in her study who were taught Monte Carlo methods without computers, showed somewhat better test results than the students with computers. Shevokas speculates about possible explanations; it may be that models had to be formulated more abstractly when used with the computer programs (p.166). Actually, such programs hardly support the modelling *process*; the user must know about the model previously, and select it from the finite choice or make the choice blindly. This seems to be different when using concrete models in a non-computer context.

Software for simulation and modelling

It is a challenge for future software development to support the construction of probability machines on the computer whilst reducing technical difficulties for the user. Although it is helpful to capitalize on physical random devices as a metaphor when using the computer (constructing spinners on a screen), this should not impede using more flexible representations with computers: the potential of mathematics to formulate and analyze hypothetical models which have no direct counterpart in the real world should also be practised by simulating artificial worlds on a computer. The best solution might be that modelling software offered various forms for expressing models: from urns to spinners to more formal representations.

It would be valuable to have more experience with software where students can design random devices on the screen. It is already possible to have an iconic representation of spinners on the screen and then express the desired model by choosing sectors. This is quite different from expressing the generation of suitable random numbers for a problem by some lines of basic code. The possibility of designing random devices on the screen is included in the software *Stochastik 1*, but its modelling capability is extremely limited, because the programs have been primarily designed to illustrate processes.

The *Logo*-microworld *Treediagram* permits a direct construction of tree diagrams on the screen via a basic command; see Dreyfus (1986) for an empirical study of using this microworld for enabling students to think more abstractly about random situations. The program *Rollerball* has an iconic interface, where the user can construct channel systems on the screen from elements such as bifurcations, input and output slots etc. In the next step, throwing balls through the system can be simulated. This should foster a structural interpretation of tree diagrams instead of only regarding them as computational tools. Its modelling and data analytic capabilities, however, are extremely limited because only bifurcation is possible with the same composition of probabilities.

The development of a more general software tool which covers the whole range of elementary probability modelling, would be a very valuable project. The kind of iconic interface interaction offered by the professional modelling system *Stella*, may serve as a prototype. It is also necessary to analyze the semantic structure which is necessary for probability modelling. Such a software should support

- the definition of elementary models/machines such as spinners, urns and boxes with several types of elements, the number of which can be specified by the user,
- the combination of elementary models/machines into complex ones with the possibility of defining the relevant outcomes,
- options for further analysis of the sampled results,
- options for stopping the sampling after a certain outcome has occurred.

The second requirement covers situations like throwing 3 dice independently with a triple of results or their sum as the relevant outcome, defining dependent experiments by 'if the outcome of the first experiment is 0, then take spinner 1, otherwise spinner 2'; or complex urn models which are used to simulate the 'game of life' etc. The *Computer Based Curriculum for Probability and Statistics* project aims at developing an easy to use software tool, its *Probability Simulator* is in this spirit (Konold et al., 1989).

However, extensive exploration of models and problems is rarely supported by existing software, which is primarily designed for demonstration rather than as a flexible tool. But, modelling tools and specific simulation programs with prepared models and options of data display and analysis can be used complementarily. In *Advanced Level Studies: Statistics*, for instance, several waiting system models are available to be compared and explored. This is a simulation for planning purposes and theoretical comparisons. In the software *Kans*, slot machines are simulated and the user can observe gains and losses. The demonstration mode is supplemented by possibilities of changing the rules of the machines and exploring the consequences. Both software examples allow the simulation to be used to solve an (imagined) problem outside the simulation. Such a model-conscious application underlines the tool character of simulations, contrary to their use as an end in themselves.

In general, solving problems by physical simulation, by computer simulation or by sophisticated mathematical and combinatorial methods, should be regarded as different yet complementary experiences for students. Non-computer simulation may be a good starting point.

Computer generated random numbers

Many reservations are related to the use of computer-generated pseudo-random numbers. Adequate perspectives may improve the use of random numbers in secondary probability education.

What distinguishes the computer as a generator of random numbers from other generators like spinners? Usually, dice and spinners are, in the first place, not considered as generators of numbers; rather, their physical properties such as symmetry and homogeneity give additional meaning to the concept of equal chances. But, with a black box random generator on a computer, there is no direct link to the physical symmetry or some other physical feature that suggests a chance mechanism. The analysis of data - which is highly theory dependent - plays the central role in judging the quality of the generator. The difficulties with pseudo-random numbers in computer simulations really are an indication of inherent features of probability; its theoretical nature is discussed in Chapter 5.

The similarity between a random generator and an urn lies in the structure of the generated data. Justifying pseudo-random number generators highlights the fact that random numbers have much more structure than just stable relative frequencies. Therefore, in the educational literature, a cautious use of computer generators at a later stage is often recommended. The value of experience with physical simulation as a prerequisite for a metaphorical understanding of computer generators is emphasized. The use of tables with random numbers is often recommended as an intermediate step.

Instead of explaining the computer generator by physical generators, the direction can be reversed: the computer generator may give additional explanation and insight into the patterns of data generated by physical generators. Integrating the computer generator into the 'zoo' of random devices from the very beginning might thus facilitate an evolution of mutually related understanding. Computer generators explain physical generators and vice versa.

From this perspective, we can utilize the many short programs for generating random sequences of numbers or of visual or geometric objects on the computer screen as is done in Watson (1980), *Mathdisk 1, 132 Short Programs for the Mathematical Classroom*, or in *Probability and Statistics Programs*. In a sense, such programs are a reaction to the challenge of random numbers which can aid understanding. This approach can be contrasted to the conservative reaction of a continued suppression of those complex mathematical phenomena that cannot be explained theoretically. It is also distinct from a theoretical approach to random numbers by explaining algorithms for their generation and techniques to test their 'randomness'. Obviously, understanding that random numbers can be imitated by a mathematical algorithm, would help to overcome the mystique surrounding computer generators, although there is still the philosophical conflict that a computer calculates a 'random' sequence deterministically.

4. Simulation and Data Analysis for Providing an Empirical Background for Probability

Using simulation for providing an empirical background is not bound to computers. The Galton Board is an early historical example for this pedagogical approach which has recently been revived in probability education. Galton Boards are supposed to have a dual function in learning probability: they may function as a real physical system or as an embodiment of an abstract chance environment, thereby indirectly simulating various physical systems. The same dual function is true of other random generators like urns or spinners and also of computer-implemented models.

Making theoretical objects experiential

Computer simulation is sometimes criticized for replacing real experience by artificial worlds; DiSessa (1986) argues against this criticism by reference to *Dynaturtle*, a computer environment for Newtonian mechanics. His basic argument is that physics is not directly about the physical world as we naturally perceive it, the world of sensory experience is not Newtonian. On the contrary, physics is about

> "abstractions that have been put together with great effort over hundreds of years (and) which happen to be very powerful once we have learned to interpret the world in their terms." (p.210)

The *Dynaturtle* is more of an Newtonian object than any real system. It is not a cheap substitute for the real thing, its significance lies in its difference from real systems. It is a temporary replacement by an abstract theoretical object: an attempt at making a theoretical object experiential, where the new possibilities of technological representation and interaction are exploited. Making something experiential is seen as a strategy for transforming naive conceptions which have proved difficult to affect by traditional ways of teaching. DiSessa stresses that the *Dynaturtle* cannot be the whole story and that a direct application of theoretical knowledge to real systems cannot be expected, but interacting with experiential software may be an important step, especially for developing qualitative knowledge.

Interestingly, in probability education we find a comparable discussion on misconceptions and primary intuitions including a diagnosis of similar difficulties with traditional teaching. Similar to the *Dynaturtle*, the computer can be programmed as an idealized chance environment with much quicker interactive and experimental opportunities than any real system. There is some evidence that experience through simulation might help; possibilities include qualitative knowledge on the role of sample size, usual variation in small samples ('law of small numbers'), the role of independence and predictability, and the idea of embedding a set of experimental data in a series of hypothetical repetitions.

Using simulation to enable students to experience an abstract chance environment instead of analyzing a given finite set of real data can have several advantages. Students participate in the process of data generation by making predictions and hypotheses, and also experience uncertainty. Moreover, experiments can be prolonged and repeated. This supports the frequentist interpretation of probability (repeatability under same conditions) more easily than other models.

The idea of providing experience in an abstract system without paying much attention to its relation to real systems, however, is very attractive for traditional mathematics education, where the role of observation, experimentation and measurement is not well established as compared to science education. We therefore find proposals stressing simulation with hardly any systematic role for real data (Råde, 1983). We find software that explicitly introduces itself as an efficient substitute for Galton Boards and other physical random generators. The rare examples of real data are simply replaced by their pseudo-random counterparts. No discussion is presented on the advantages of real and simulated data.

Many papers which discuss the role of simulation do not really consider whether using real data is more appropriate, or is an essential extension for the objective they have in mind, but this attitude should become standard. One of the obstacles, namely that access to large data bases and flexible software for data processing and analysis cannot be assumed to be present in schools, will probably disappear in the future. But there are also deeper, more philosophical obstacles, which will be discussed with several examples.

Beginning with 'limited' technological equipment

Råde (1977; 1981; 1983) pioneered the use of computer simulation in probability education. He has the same background as Engel. Engel (1986) still argues in favour of using (*BASIC*-programmable) hand held calculators in probability education. The main advantage is that each student has hands-on access to the machine - which is, at present, difficult with micros.

Reading such early papers highlights the fundamental challenge raised by the novel technology, that is the possibility of choosing and adapting the system of primitives in pedagogical situations. With the hand held calculator without a *RND*-command, random numbers have to be generated first. Generating different kinds of probability distributions while taking the *RND* for granted as a primitive was the next step. Only this bottom-up approach was possible. In current educational software, much more elaborate primitives are taken for granted. Råde (1983, pp.32) gives a list of possible advantages or objectives of using simulations:

1) A combination of a theoretical and experimental approach may be used.
2) Important concepts like probabilities, expectation and variance and their properties can be motivated and illustrated by appropriate experiments.
3) Work with data obtained from simulations will provide many opportunities for statistical data analysis.
4) Stability and variability in random experiments and statistical estimation procedures can be illustrated.
5) The students will learn about simulation, which is an important method of solving complex probability problems.
6) Simulation of random experiments can be organized as teamwork of groups of students.

Many examples from the literature can be interpreted along these lines. The basic common idea is to use the computer as a cheap and convenient resource for all kinds of random experiments which can provide a lot of data and experience that serves as a counterpart of theoretical work.

Laws of large numbers and frequentist interpretation

Often, students understand the laws of large numbers only superficially. Without computer support, it is difficult to work with large numbers and long run frequencies remain mysterious. Many misconceptions have been found in this area; students act as if a 'law of small numbers' holds; they share the misconception that there is a stabilization of *absolute* frequencies; they do not develop a feeling about the order of variability in moderate sample sizes nor do they see an empirical meaning in important formulae and concepts. The computer can help to develop relevant ideas. For instance, the relationship between the standard deviation of the arithmetic mean and the original standard deviation ('square root *n* law') can be developed as a descriptive measure of variation in repeated series of tossing coins (Freudenthal, 1974).

This 'square root *n* law' may be linked to a twofold experience. With adequate data bases and software tools, the law of large numbers can now be discovered and described as a real regularity, not only as a regularity in simulated numbers. Collis (1983) suggests counting letter frequencies in texts for that purpose. However, it is also possible to reproduce similar patterns with artificial simulated data. Deviations from the pattern may be used as clues for a further investigation of the underlying situation.

Let us turn to some practical suggestions. A standard approach is plotting or tabulating relative or absolute frequencies against the number of trials, observing paths in time, comparing different paths of the same length etc. (*Stochastik 1*). The distribution of relative frequencies for a fixed sample size is simulated. Then the investigation of this distribution for increasing sample sizes yields the law of large numbers and the central limit theorem as a side product. There is evidence that students appreciate and understand these 'laws' much better after such simulations (Johnson, 1986).

These applications enable a deeper understanding of the laws of large numbers. However, most programs are only demonstrations with limited student activities. Extended data analysis would enable students to gain a more differentiated understanding of structures and phenomena in random sequences. An appropriate interactive computer environment which also poses these interesting challenges for exploration has not been designed yet, but the already mentioned *Probability Simulator* seems to be a step in this direction.

Another approach is to illustrate the convergence of a sample distribution towards its theoretical distribution (see *Advanced Level Studies: Statistics*). It is unlikely that, at school level, mathematical theory will ever be developed to explain this phenomenon theoretically. In this sense, simulation is not only used as an empirical basis or counterpart for a coming formal theory; rather, the computer offers new opportunities to acquire informal knowledge about probability.

Philosophically, laws of large numbers can be interpreted as the existence of statistical regularities despite irregularities and unpredictability in individual cases. The Galton

Board is an excellent embodiment of this principle. Simulation and data analysis can be used for exploring this basic law in its many facets. For example, Bar-On and Or-Bach (1988) take a two dimensional random walk in a rectangular grid of length $k+1$ and visualize the repeated process on the screen; this is mathematically equivalent to a Galton Board. In order to demonstrate that the distribution shows a predictable pattern, they ask students to investigate the influence on the shape of the distribution when the basic parameter p for the decision to walk horizontally instead of vertically is changed. The emerging pattern is a different experience to deducing the distribution from the basic assumptions of a random walk. The procedure does not provide a proof; rather, it gives an empirical explanation.

Should such computer experiments replace real Galton Boards? If one aims only at a combinatorial analysis of the paths, a simulated board on the screen may be enough, but a real Galton Board can be used to discuss the meaning of probability and randomness in real systems (Jäger and Schupp, 1983). It supplies possibilities for

- looking for explanations for the fact of the 'favourable middle'
- testing whether prediction of paths is possible
- testing whether the path of the ball can be influenced
- discussing which physical conditions may be responsible for equal chances and 'local' and 'global' independence
- using this knowledge to build simple Galton Boards
- comparing the real distribution of balls with the theoretical expectation (Steinbring, 1985)

Another step is to explore rich databases of real data. For instance, if we explore the proportion of males in birth data, a change from daily, to weekly, monthly and yearly proportions will show increasingly stable patterns.

Random sampling and sampling variation

For probability education, two aspects of random sampling are particularly relevant; how to draw a random sample from a real population and, theoretical knowledge about the behaviour of various random samples from the same population. The latter topic is at the interface between probability and (inferential) statistics. These concepts are difficult to grasp, especially the fundamental idea of embedding a single event (what has actually happened) into a system of hypothetical events (what might have happened). Before the computer age, physical devices like boxes with many beads have been used to give more life to these abstract concepts; the hypothetical series becomes more realistic, variation from sample to sample can be observed, the possibility of making wrong decisions can be experienced etc. However, an adequately programmed computer can be more powerful than a bead box for various reasons:

* It is difficult to construct reliable and efficient bead boxes.
* The content of a 'computer bead box' can be freely varied.
* It is possible to choose a box which is hidden from the learners.
* Large and varied samples are easily generated.
* Data are easily stored for further processing.

A major argument for changing from the real bead box to the computer is reducing noise and saving time for essential things; however, poor software and other difficulties with using the computers can be counterproductive. It seems reasonable to use computer bead boxes only after some basic experience of drawing random samples from real boxes; this is the pedagogical approach of *Sampling Probability and Prediction*.

Many educational packages in probability contain programs on sampling. For example, the sampling programs in *Advanced Level Studies: Statistics* are designed to enable visualization of mathematical processes; the user has some control over parameters and the form of the parent distribution. In a program on the random character of confidence intervals, the result of a sample is shown, a confidence interval is plotted, and the display shows whether the true value lies inside the interval. Repeating this process and plotting the intervals one below the other shows the sampling variation of confidence intervals. A similar kind of dynamical visualization appears in a subprogram, where the result of sampling from a user-chosen distribution is shown in a histogram changing stepwise. At the same time, the arithmetic mean is plotted in another histogram to show its sampling distribution. Some empirical evidence suggests the superiority of similar programs over static visualization (Johnson, 1986).

Merely showing sampling variation as many programs do, is not enough to achieve understanding. Several programs allow graphical and numerical exploration with various distributions. Other pieces of software present challenges, where sampling variation and sample size are a critical elements in solving problems. The program *Guess my Bag* from the collection *Sampling Probability and Prediction* asks students to find which of three possible bags (containing different distributions of four sorts of beans) has been chosen by the computer. Students can draw samples of the hidden bag and use graphical display for comparing the sample with the three possible populations, chosen randomly at the beginning. They thereby explore the role of sample size; a random generator is used to fill the bead boxes so the teacher does not know the right answer either. Although the basic idea is promising, its design, like many other pieces of educational software, is rather closed; it does not offer a spectrum of challenges or options and lacks an adequate teacher interface (Rach et al., 1988).

Other simulation software utilizes a broader context. For instance, in *Opoll*, students can draw samples and are asked to predict which political party will win an election; or, the user can run a market research firm where the quality of predictions has financial consequences. Thus political or financial problems intermingle with sampling issues. This is even more the case in the program *Prime Time* from the software *Sampling: Probability and Prediction*.

A tool-like program for studying sampling variation is *StatLab*. It is much more comprehensive than the above mentioned programs. The focus is on laboratory experiments which enable the students to study the statistical concepts of an introductory statistics course by observing how they vary under repeated sampling.

Structure in random sequences

Some programs do not focus on the sampling distribution but on the structure of a sequence of random data. Can we distinguish a sequence of numbers emerging from tossing coins from another which somebody has just written down on a sheet of paper? Randomness is highly structured compared to haphazardness. The assumption of independent stochastical trials has many empirical consequences such as the number of pairs, triples etc. in a sequence. Elementary knowledge about these phenomena is important for appreciating random numbers and their generators. It is also well known that students have difficulties with such features of randomness. Computers can provide empirical background for the necessary conceptual development.

One approach consists in simulating the throw of a coin and plotting the sequence of results with two different symbols on a computer screen. Changing the basic probability p of a head will have well known visual effects. Green has developed a collection of programs for attacking this problem (*Probability and Statistics Programs*). We will briefly look at the program *C.I.A.* (*Cointosser Invention Analyzer*), which seems to be the most instructive and convincing of the collection.

In addition to displaying data visually, *C.I.A.* includes methods to analyze data. Its basic idea consists of concealing nine 'coin machines' in the computer which produce sequences of binary data. Each machine is either a fair coin tosser or represents some kind of deviation from the model of equal chance and independence. Such deviations are: the probability *p* differs from 1/2; after head another head is more likely; after head a tail is more likely; heads mostly come first, tails at end; a repeated pattern; a pattern is randomly chosen. Several commands for graphical and numerical analysis of the data are provided. Data can be grouped (group size 2 - 10) and the distribution plotted or tabulated. The number of runs as well as the distribution of run length can be calculated.

In contrast to many other programs, the simulated data are built up and presented on the screen and can then be analyzed from several points of view. In this rich and moderately complex environment, students can experiment with the machines, make interactive data analyses, model or describe their behaviour, apply and develop theoretical knowledge about binomial distributions or about runs, decide whether a machine seems to be fair or not, etc. According to the accompanying teaching material, a major aim is that students learn to discriminate between a fair coin tosser and an unfair one. However, the environment can be also used for slightly different pedagogical purposes. A basic option is the possibility of repeating an experiment if the user is not sure how to judge a machine. Obviously, there will be individual differences in judgement which can stimulate discussion. Such experience in an ideal chance environment may also be used to provide valuable experience for understanding statistical tests and applying them reasonably. Textbooks on statistical inference only give tests as being definitive and do not explain that though uncertainty is inevitable it may be reduced by further experiments.

Another option is to input data from the keyboard, for instance from real coin tossing experiments or from students' attempts to generate random sequences by hand. This option for combining simulation with real data analysis, however, is only modestly supported; real data cannot be stored and retrieved from discs. The accompanying material discusses the problem of relating the experience in the learning environment to analyzing structures in sequences of real data and interpreting deviations from the ideal pattern in terms of the real situation. Such experience would be a further step in understanding randomness in the sense Feller (1968) describes:

> "In testing randomness, the problem is to decide whether a given observation is attributable to chance or whether a search for assignable causes is indicated ..."

Related to runs he discusses examples

> "...counting runs of boys and girls in a classroom might disclose the mixing to be better or worse than random. Improbable arrangements give clues to assignable causes; an excess of runs points to intentional mixing, a paucity of runs to intentional clustering." (p.42)

In summary, experience in an abstract chance environments is valuable and difficult to achieve without a computer. But using such knowledge and experience as a reference and tool for exploring real data and systems in a second step is the ultimate goal of understanding randomness.

A simulation and modelling tool as companion of the curriculum

The *Computer Based Curriculum for Probability and Statistics* project aims at software that can accompany a curriculum where the empirical side of probability is an integrated feature (Konold et al., 1989). This comprehensive tool provides simulation data for the above mentioned problems and for some more which have been identified by research. The central idea is that students express their beliefs about the probability of events. The software allows one to define a model and sample from it to test and change these beliefs. As the student activity is central, this software is quite different from others which merely demonstrate something by simulation.

Some of the problems of the pilot curriculum material are the following (Konold, 1991). In *Coin Flipping*, students are asked to empirically explore the probability of patterns like *HHTHT*, or the probability of *H* as compared to *T* after a series of *H*'s. This is directed towards overcoming the gambler's fallacy and the problems with the representativeness heuristics (see Chapter 7). In *Random vs Mixed Up*, students can compare coin flipping to artificial data; in *Coincidences*, a box model for the birthday problem is defined and can be analyzed with regard to coincidences.

The software tool *Probability Simulator* provides a well organized system of commands which enable the student to analyze diverse simulated data. For instance, it is possible to search for specified patterns in a sequence of numbers and count the number of occurrences, to count the (conditional) frequency of outcomes after a certain pattern, to draw a sample until a certain event has occurred for the first time etc. The system of commands is adapted to such problems and therefore different from those commands which are usually available with standard statistical software.

Games and strategy

In computer assisted instruction, games are often related to a topic rather superficially. Obviously, this is different in probability where they provide a major source for the development of probability theory and are still an important domain of application. Moreover, the statistician's role has been defined as 'playing games against nature'.

Games of chance have often been exploited as a method of teaching and learning. Can computer-implemented games offer additional opportunities for learning? O'Shea and Self (1983) express a

> "desperate need for experimental studies of their [computer games] educational effectiveness."

Existing studies support the sceptics and are consistent with the critical evaluation of the disparity of computer experience that Bauersfeld (1984) has elaborated.

> "In spite of motivationally attractive 'packaging', the microcomputer games used in this study were not very effective at teaching probability and estimation. Given the proven success of non-computer games, this result was somewhat surprising and raises the possibility that students may not process information presented in a computer environment in the same way that they process information in a non-computer environment. Much more study is needed of appropriate instructional uses of computers so that teachers know how to exploit the best features of such environments." (Bright, 1985, p.522)

In the following, a brief review of some pieces of software and related ideas on games is given. *Subgame* is part of *Micros in the Mathematics Classroom.* Digits are randomly generated and the students have to decide the place they should be put in the subtraction for a maximal result. Some empirical research on the use of this software from the *ITMA* Project is reported by Fraser et al. (1988, pp.330). The influence of the teacher and the changing of classroom roles are important messages of those studies; it seems difficult to jump from playing the game to reflecting on strategies and the stochastic characteristics of the game.

The program *Strategy* involves a similar task. The main difference consists in additional options for defining a strategy. Different strategies can be programmed in a specific language and improved and compared empirically. This is an option that goes far beyond the possibilities of non-computer games. This feature of exploring competing strategies is also present in other software, such as *Probability Games*, which, however, only makes limited use of the flexibility of computers.

Simulation can be used to broaden the perspective of an individual player, as found in the program *Kans*. In a roulette simulation, the user can simultaneously observe gains and losses of the bank and of about 10 players in a dynamic graphical display. He/She can give each computer player an initial amount of money, and then decide the combination of numbers to play. Interesting experience can be gained in such a modifiable environment; in a simulation of one-armed bandits, students can assume the designer's role.

Last but not least, computer analysis of real data on public games of chances may be very instructive, especially if data on biased lotteries or games are available; e.g., in the military draft lottery in the *USA* severe deviations from randomness occurred (Moore, 1990, p.133). An analysis of lottery numbers people actually used and the related lottery pay-off is an instructive introductory example in a book on interactive data analysis with the statistical language *S* (Becker et al., 1988).

5. Visualization, Graphical Methods and Animation

Statistics has experienced a revolutionary use of interactive graphical methods for data analysis. In the simulation of complex systems, multiple representations with different levels of abstraction are used. For instance, a flight simulator provides an analogical video-like representation of an environment. This diversity is only partly reflected in current software for the classroom. Most software has intuitive appeal but is not supported by a substantive theory of visualization. We briefly summarize the potential by a preliminary classification from Lunn (1985):

(1) Graphs for communicating and exploring real data or data from simulations.

(2) Graphs used as general mathematical utilities, for studying functional relations, solving equations approximately, illustrating and exploring concepts (e.g. binomial distributions).

(3) Visualizing random patterns such as sequences of random numbers, Poisson distribution in the plane.

(4) Visualizing stochastic processes such as sampling; laws of large number; central limit theorem.

(5) Visualizing mathematical processes such as the process of generating Pascal's triangle or other computational algorithms.

(6) Schematic or iconic representation of 'real systems' as a modelling interface or as a cognitive and motivational support for prepared simulations.

(7) Exterior animation, which is only superficially related to the subject matter but used for pedagogical purposes.

Using computers as a visualizing tool is the main focus of a project which develops the software *Elastic*. The main principles are (see Swets et al., 1987, p.9):

1. The computer shall be used as a visualization tool. Since many aspects of probability can be more effectively understood if they are illustrated graphically, the computer adds new dimensions and the possibility to see more pictures and dynamical representations.
2. Inclusion of multiple, linked representations in the sense of Kaput (1986); the theoretical assumption is that different students may find different representations persuasive and memorable, and that all students will understand concepts more deeply if they comprehend the connections between representations.
3. Providing opportunities for interaction. Students can interact directly with statistical objects. This interaction takes place in a mode different from the standard question--and-answer format. Students can modify data and graphs, set parameters for sampling experiments, and construct decision models. The immediate feedback provides students with the power to 'discover' statistical concepts.

The principle of multiple, linked representations, however, has to be counterbalanced with the fact that each type of representation requires a learning effort and that the multitude might confuse. The above pedagogical perspective on visualizations has to be theoretically supplemented by a perspective on graphs as professional and cultural tools. From this perspective, students need support to develop their skills in using a selected set of important graphs. For instance, the tree diagram is one of the most important graphical tools for probability. In this spirit, the *ITMA* project has produced computer--supported material for the *Language of Functions and Graphs*. Fraser (1989) elaborates a related developmental approach to 'the vocabulary of modelling' including graphs and diagrams. A similar perspective on graphs in *probability* is missing, although there are attempts to restructure *statistics* from this point of view, taking into account the revolutionary use of statistical graphs in practice.

The dangers of motivation by audio-visual effects which are only superficially related to the subject matter have often been debated. In *Stochastik 1*, a Galton Board is simulated as non-interactive video. Little drunk people, walk randomly through a forest where the trees are arranged like the pins in a Galton Board, at the end they fall into a certain part of a hotel. The *Mime Statistics* series also makes extensive use of animation; we find exterior animation in pure terms. For instance, students mark a place in a histogram where they think the median of a distribution lies. The whole histogram is then filled with water and split into two parts according to the mark; a water network is filled with water, so that a little person gets wet feet if the two water parts are not equal.

Many similar analogous representations deserve deeper exploration with regard to their educational value. Examples are analogous representations of a doctor's waiting room

(*Advanced Level Studies: Statistics*); traffic flow; Eskimo fishing; and a shooting situation in *Mime Statistics*; dice throwing and drawing balls from a bag in *Probability Games*. The software *GASP* which is designed for visualizing stochastic processes at college level also has many interesting visualizations.

Such types of representations have always played a role in probability education. There is an important difference with representations on a computer; the screen is usually only the surface representation which is controlled by the deeper representation, namely the mathematical or computational model which is programmed. Analogous representations of stochastic systems whose model is implemented and simulated on the computer can have various different functions: they should be judged according to whether they reasonably represent the semantics of a model, illuminate the modelling assumptions and can open a new world of experience which otherwise cannot be reached efficiently.

6. Concluding Remarks

It is clear that deficiencies on several levels have to be addressed by empirical research in order to deepen theoretical perspectives on educational computing in probability. Simulation and modelling software for secondary probability education that combines ease of use with the flexibility and adaptability of programming languages seems not to exist at present. Such a tool is needed if the potential of an extended modelling approach is to be realized. There are some requirements which should become the object of further empirical and theoretical research:

- supporting the process of modelling through a construction set which allows modelling on several levels of abstractness, for instance, operating with iconically represented chance environments as well as with tree diagrams and more formal representations;
- flexible options for selecting or defining the random variables of interest (waiting times, events, distributions etc.);
- control over a wide range of display and analysis facilities for studying the generated data of interest from several points of view;
- possibilities for input of real data or simulated data from other sources should be available; options for comparing models, simulated data, and real data should be available at several levels of sophistication.

Professional software for dynamic modelling *(DMS, Stella,* spreadsheets), for interactive statistical analysis (*StatView, Data Desk*) and interactive languages like *APL, PC-ISP* and *S* should be further explored, with the goal of specifying basic requirements of an educational tool. A more detailed specification needs to be developed of what such software should do in terms of tasks and in terms of the knowledge and competence required of the user. Current programs intended to attack the 'lack of experience' are not flexible enough and the developmental attempts are not linked to research findings about students' difficulties. The *Probability Simulator* seems to be an exception to this rule. Some pieces which offer a certain flexibility and range of options, however, could be quite successful.

This chapter has concentrated on the subject matter from an epistemological point of view, on new styles of work, on tools and representations as well as on possible changes in contents and on a comparison to non-computer environments. The central focus has been on modelling and simulation. As this book is on probability, statistical data analysis and inferential statistics were mentioned only briefly. Nonetheless, this separation is artificial; computers will help to emphasize the statistical aspects of probability. Moreover, if we look at the possible impact of computers on the probability and statistics curriculum as a whole, there is a transformation of descriptive statistics in the direction of Exploratory Data Analysis. Learning and applying probability may profit from more extended students' experience in these domains as an empirical basis, even when data analysis is of more limited relevance to the probability part of a curriculum. A new balance and relation of these aspects has to be still further explored from both the probabilistic and the statistical perspective.

Software

132 Short Programs in the Mathematical Classroom, Higgo: 1985, Stanley, Thornes & Hulton, Old Station Drive, Leckhampton, Cheltenham, GL53 ODN, England, ISBN 0 85950 556 4, BBC computer.

Advanced Level Studies: Statistics, V. Barnett, P. Holmes et al.: 1984, BBC Publications, 35 Marlebone High Street, London W1M 4AA, England, BBC computer.

Data Desk 3.0, P.F. and A.Y. Velleman: 1989, Odesta Corporation Inc., 4084 Commercial Avenue, Northbrook, ILL 60062, APPLE MacIntosh.

Derive V.2: 1990, Soft Warehouse, Inc., 3615 Harding Ave, Suite 505, Honolulu, Hawai 96816-3735.

DMS (Dynamic Modelling System), J. Ogborn: 1985, Longmann Micro Software, 62 Hallfield Unit, Layerthorpe, York, Y03 7XQ, England, BBC/APPLE.

Elastic (pre-release version): 1989, Bolt, Beranek & Newman, Cambridge (Mass.), APPLE MacIntosh.

GASP Graphical Aids for Stochastic Processes, Fisch, B. and Griffeath, D.: 1987, Wadsworth and Brooks/Cole Advanced Books & Software, IBM compatible.

Kans (pre-release version), P.v. Blokland: 1989, 1103 RK Amsterdam, Geerdinkhof 561, IBM compatible.

MathCAD, 2.0: 1988, The Engineers Scratch Pad. Math Soft, 201 Broadway, Cambridge MA 023139, IBM compatible.

Mathdisk 1, C. Kimberling: 1984, University of Evansville Press, P.O. Box 329, Evansville, IN 97702 USA, ISBN 0 930 982 12 6, APPLE II.

mathematica: 1988, see Wolfram, S., *Mathematica*, Addison Wesley, Reading, Apple MacIntosh and IBM compatible.

Micros in the Mathematics Classroom, Shell Centre for Mathematical Education: 1984, Longman Group Ltd., Longman House, Burnt Hill, Harlow, Essex, England, ISBN = 582 26510 x, BBC.

Mime Statistics 1 - 5, A.C. Bajpai et al.: 1987, J. Wiley, BBC computer.

Opoll: 1985/86, Muse, P.O. Box 43, Hull HU1 24D, England, BBC computer.

PC-ISP, Artemis Systems Inc., 1985, PC ISP-Interactive Scientific Processor, Champman & Hall, New York/London, IBM compatible.

Probability Games, 1986, Encyclopaedia Britanica Educational Corporation, 425 North Michigan Avenue, Chicago, Illinois 60 611, USA, ISBN 0 8347 3344 7, APPLE II.

Probability and Statistics Programs, Green, D. et al.: 1986, Capital Media, ILECC, John Ruskin Street, London SE5 OP2, ISBN 0 7085 9892, BBC computer.

Probability Simulator (software in development), Konold, C. et al: 1991 (ongoing), University Massachusetts, Amherst, Scientific Reasoning Research Institute, Apple Macintosh.

Resampling Stats: 1989, 612 N. Jackson St., Arlington, VA 22201, IBM compatible.

Rollerball: 1986, Program in the collection *Probability and Statistics Programs.*

S, see Becker et al. (1988).

Sampling Distributions, G. Alderson, and M. Dodwell: 1984, Longman House, Burnt Mill, Harlow, Essex CM20 2JE, ISBN APPLE Disk 0 582 26744 7, BBC, 0 582 26747 1, APPLE II, BBC computer.

Sampling: Probability and Prediction, W.F. Finzer et al.: 1986 D.C. Heath & Company, 125 Spring Street, Lexington, Massachusetts 023173, USA, ISBN 0 669 12221 1, APPLE.

SPSS/PC+, SPSS Inc, 444 N. Michigan Ave, Chicago IL 60611, IBM compatible.

StatView, Se+ Graphics, 1988, Abacus Concepts, Inc., 1984 Bonita Avenue, Berkeley CA 94704, APPLE MacIntosh.

StatLab, D.W. Stirling: 1987, The New Zealand Statistical Association, Wellington, P.O. Box 1731, Apple Macintosh.

Stella: 1986, High Performance Systems, 13 Dartmouth College Highway RR1, Box 37, Lyme NH 03768, USA, APPLE MacIntosh.

Stochastik 1,2, Steller, E. et al.: 1984, KLETT, ISBN 3 12 11070 X, 3 12 110090 7, APPLE II.

Strategy: 1986, F. Finzer and D. Resek, Menlo Park Addison-Wesley, Apple II.

Tree Diagram, P.W. Thompson: 1984, Cosine Inc., P.O. Box 2017, W. Lafayette, IN 47906, USA, Commodore C64.

Bibliography

Alvord, L.L.: 1985, *Probability in APL: A Model for Using Executable Language*, Ph.D.thesis, Rutgers University, New Brunswick.

Atkinson, D.T.: 1975, *A Comparison of the Teaching of Statistical Inference by Monte Carlo and Analytic methods*, Ph.D.thesis, University of Illinois, Urbana Champaign.

Bar-On, E. and R. Or-Bach: 1988, 'Programming Mathematics: A New Approach in Introducing Probability to Less Able Pupils', *International Journal for Mathematical Education in Science and Technology* **19**, 281-297.

Bauersfeld, H.: 1984, 'The Disparity of Computer Experience - a Case for Orienting the Syllabus for Elementary Education?', in J.D. Tinsley (ed.), *Informatics in Elementary Education*, North Holland, Amsterdam, 199-206.

Becker, R.A., J.M. Chambers and A.R. Wilks: 1988, *The New S Language*, Wadsworth and Brooks, Pacific Grove.

Biehler, R.: 1982, *Explorative Datenanalyse - Eine Untersuchung aus der Perspektive einer deskriptiv - empirischen Wissenschaftstheorie*, Materialien und Studien 24, IDM, Bielefeld.

Biehler, R.: 1985, 'Interrelations between Computers, Statistics and Teaching Mathematics', in Commission Internationale de L' Enseignement Mathematique (ed.), *The Influence of Computers and Informatics on Mathematics and Its Teaching, Supporting Papers*, IREM, Strasbourg, 209-214.

Biehler, R.: 1991, 'Fortschritte der Software und die Tradition der Schulmathematik', in W. Dörfler, W. Peschek, and E. Schneider (eds.), *Mensch-Computer-Mathematik*, Hölder-Pichler-Tempsky, Wien.

Biehler, R. and W. Rach: 1990, *Softwaretools zur Statistik und Datenanalyse: Beispiele, Anwendungen und Konzepte aus didaktischer Sicht*, Soester Verlagskontor, Soest.

Bright, G.W.: 1985, 'What Research Says: Teaching Probability and Estimation of Length and Angle Measurement through Microcomputer Instructional Games', *School Science and Mathematics* **85**, 513-522.

Collis, B.: 1983, 'Teaching Descriptive and Inferential Statistics Using a Classroom Microcomputer', *Mathematics Teacher* **76**(5), 318-322,

diSessa, A.: 1986, 'Artificial Worlds and Real Experience', *Instructional Science* **14**, 207-227.

Dreyfus, T.: 1986, 'Cognitive Effects of Microworlds: Learning about Probability', *Proceedings Second Intern. Conf. Logo and Math. Educ.*, London, 172-179.

Efron, B.: 1979, 'Computers and the Theory of Statistics: Thinking the Unthinkable', *SIAM Review* **21**, 460-480.

Engel, A.: 1975, 'Computing and Probability', in L. Råde (ed.), *Statistics at School Level, Proc. Third ISI Round Table Conference on the Teaching of Statistics*, Wiley, New York, 95-120.

Engel, A.: 1977, *Elementarmathematik vom algorithmischen Standpunkt*, Klett, Stuttgart.

Engel, A.: 1985, 'Algorithmic Aspects of Stochastics', in Commission Internationale de L' Enseignement Mathematique (ed.), *The Influence of Computers and Informatics on Mathematics and Its Teaching, Supporting Papers*, IREM, Strasbourg, 215-222.

Engel, A.: 1986, 'Statistik mit programmierbaren Taschenrechnern (PTR) und Tischrechnern', *Didaktik der Mathematik* **14**, 243-263.

Engel, A.: 1987, *Stochastik*, Klett, Stuttgart.

Feller, W.: 1968, *An Introduction to Probability Theory and its Applications*, Wiley, New York.

Fraser, R.: 1989, 'Role of Computers for Application and Modelling. The ITMA Approach', in W. Blum, J. Berry, R. Biehler, I. Huntley, G. Kaiser-Messmer and L. Profke (eds.), *Applications and Modelling in Learning and Teaching Mathematics*, Ellis Horwood, Chichester, 373-380.

Fraser, R., H. Burkhardt, J. Coupland, R. Phillips, D. Pimm and J. Ridgway: 1988, 'Learning Activities and Classroom Roles with and without Computers', *Journal of Mathematical Behavior* **6**, 305-338.

Freiberger, W. and U. Grenander: 1971, *A Course in Computational Probability and Statistics*, Springer, New York.

Freudenthal, H.: 1974, 'The Crux of Course Design in Probability', *Educational Studies in Mathematics* **5**, 261-277.

Freudenthal, H.: 1975, *Wahrscheinlichkeit und Statistik* 3rd. ed., Oldenbourg, München.

Gnanadesikan, M., R.L. Schaeffer, and J. Swift: 1987, *The Art and Technique of Simulation*, Dale Seymour, Palo Alto.

Harten, G. von, H.N. Jahnke, T. Mormann, M. Otte, F. Seeger, H. Steinbring, and H. Stellmacher: 1986, *Funktionsbegriff und funktionales Denken*, Aulis, Köln.

Holmes, A. H.: 1968, *Teaching the Logic of Statistical Inference by the Monte Carlo Approach*, Ph.D.thesis, Univ. Illinois, Urbana.

Holmes, P.: 1985, 'Using Micro Computers to Extend and Supplement Existing Material for Teaching Statistics', in L. Råde and T. Speed (eds.), *Teaching of Statistics in the Computer Age. Proceedings of the 6th ISI Round Table Conference on Teaching of Statistics,* Chartwell Bratt Ltd., Bromley, 87-104.

Howson, A.G. and J.P. Kahane (eds.): 1986, *The Influence of Computers and Informatics on Mathematics and Its Teaching*, Cambridge Univ. Press, Cambridge.

Inhelder, W. : 1981, 'Solving Probability Problems through Computer Simulation', in A.P. Shulte and J.R. Smart (eds.), *Teaching Statistics and Probability*, NCTM Yearbook 1981, National Council of Teachers of Mathematics, Reston VA, 220--234.

Jäger, J. and H. Schupp: 1983, *Curriculum Stochastik in der Hauptschule*, Schöningh, Paderborn.

Johnson, N.C.: 1986, *Using a Microcomputer to Teach a Statistical Concept*, Ph.D.thesis, Univ. Minnesota.

Kaput, J.J.: 1986, 'Information Technology and Mathematics. Opening New Representational Windows', *Journal of Mathematical Behavior* 5, 187-207.

Kaput, J.J.: 1988, *Looking Back From the Future. A History of Computers in Mathematics Education, 1978-1998*, Unpublished Manuscript, Educational Technology Center, Cambridge MA.

Klingen, L.H.: 1977, 'Zusammenhang und Reichweite modularer Algorithmen in der Schulmathematik', in *Informatik im Unterricht der Sekundarstufe II. Grundfragen, Probleme und Tendenzen mit Bezug auf allgemeinbildende und berufsqualifizierende Ausbildungsgänge*, Schriftenreihe des IDM 16, IDM, Bielefeld, 3-33.

Konold, C.: 1991, *One Son, Coincidences, Coin Flipping, Random vs. Mixed up, Building Box Models*, Pilot student materials from the Computer Based Curriculum for Probability and Statistics Project, Univ. Massachusetts, Amherst MA.

Konold, C., M. Sutherland, and J. Lochhead: 1989, *A Computer Based Curriculum for Probability and Statistics*, Project Proposal, Univ. Massachusetts, Amherst MA.

Landwehr, J.M., J. Swift, and A.E. Watkins: 1987, *Exploring Surveys and Information from Samples*, Dale Seymour, Palo Alto.

Lunn, D.: 1985, 'Computer Animation: A Powerful Way of Teaching Concept of Probability and Statistics', in L. Råde and T. Speed (eds.), *Teaching of Statistics in the Computer Age. Proceedings of the 6th ISI Round Table Conference on Teaching of Statistics,* Chartwell Bratt Ltd., Bromley, 114-125.

Moore, D.S.: 1990, 'Uncertainty', in L.A. Steen (ed.), *On the Shoulders of Giants*, National Academy Press, Washington D.C., 95-137.

O'Shea, T. and J. Self: 1983, *Learning and Teaching with Computers*, Harvester, Brighton.

Pollak, H.O.: 1989, 'Recent Applications of Mathematics and their Relevance for Teaching', in W. Blum, J. Berry, R. Biehler, I. Huntley, G. Kaiser-Messmer and L. Profke (eds.), *Applications and Modelling in Learning and Teaching Mathematics*, Ellis Horwood, Chichester, 32-36.

Rach, W., R. Biehler, and B. Winkelmann: 1988, 'Teacher Interfaces in Software for Mathematics Education: Different Requirements of Software from the Perspectives of Learners and Teachers', in A. Vermandel (ed.), *Theory of Mathematics Education. Proceedings of the Third International Conference*, Univ. Antwerpen, Antwerpen, 149-156.

Råde, L.: 1977, 'Probability, Simulation, and Programmable Calculators', in W. Dörfler and R. Fischer (eds.), *Anwendungsorientierte Mathematik in der Sek. II*, Hölder-Pichler-Tempsky, Wien, 163-177.

Råde, L.: 1981, 'Statistics, Simulation, and Statistical Data Analysis at the School Level', in W. Dörfler and R. Fischer (eds.), *Stochastik im Schulunterricht*, Hölder-Pichler-Tempsky, Wien, 165-178.

Råde , L.: 1981a, 'Random Digits and the Programmable Calculator', in A.P. Shulte and J.R. Smart (eds.), *Teaching Statistics and Probability*, NCTM Yearbook 1981, National Council of Teachers of Mathematics, Reston, 118-125.

Råde, L.: 1983, 'Stochastics at the School Level in the Age of Computer', in D.R. Grey, P. Holmes, V. Barnett, and G.M. Constable (eds.), *Proceedings First Intern. Conf. Teach. Statistics vol. 1*, Teaching Statistics Trust, Sheffield, 19-33.

Riemer, W.: 1985, *Neue Ideen zur Stochastik*, Bibliographisches Institut, Mannheim.

Sanders, W.J.: 1971, *Teaching Statistical Decision Making to Junior High School Students*, Ph.D.diss, Univ. Illinois.

Shevokas, C.: 1974, *Using a Computer-Oriented Monte Carlo Approach to Teach Probability and Statistics in a Community College General Mathematics Course*, Ph.D.diss, Univ. Illinois.

Simon, J.L., D.T. Atkinson, and C. Shevokas: 1976, 'Probability and Statistics. Experimental Results of a Radically Different Teaching Method', *American Mathematical Monthly*, **83**, 733-739.

Simon, J.L. and D. Weidenfeld: 1973, *Simple. A Radical Teaching and Computer Program for Statistics* , Univ. Illinois.

Steinbring, H.: 1985, 'Wie verteilen sich die Kugeln beim Galton-Brett wirklich?', *mathematik lehren* **12**, 31-38.

Swets, J.A., A. Rubin, and W. Feuerzeig: 1987, *Cognition, Computers, and Statistics. Software Tools for Curriculum Design*, Progress Report-Year 1, BBN Laboratories, Cambridge.

Tall, D. and B. Winkelmann: 1988, 'Plotting Function Graphs and Teaching Hidden Algorithms', *Bulletin of the I.M.A.* 24(7/8), 111-115.

Travers, K.: 1981, 'Using Monte Carlo Methods to Teach Probability and Statistics', in A.P. Shulte and J.R. Smart (eds.), *Teaching Statistics and Probability*, NCTM Yearbook 1981, National Council of Teachers of Mathematics, Reston, 210-219.

Travers, K.J., W.F. Stout, J.H. Swift, and J. Sextro: 1985, *Using Statistics*, Addison Wesley, Menlo Park.

Tukey, J.W.: 1962, 'The Future of Data Analysis', *Annals of Mathematical Statistics* 33, 1-67.

Watkins, A.E.: 1981, 'Monte Carlo Simulation: Probability the Easy Way', in A.P. Shulte and J.R. Smart (eds.), *Teaching Statistics and Probability*, NCTM Yearbook 1981, National Council of Teachers of Mathematics, Reston, 203-209.

Watson, F.R.: 1980, *A Simple Introduction to Simulation. Part I+II*, Keele Mathematical Education Publications, Keele.

Rolf Biehler
Institut für Didaktik der Mathematik
Universität Bielefeld
Postfach 8640
DW - 4800 Bielefeld
F.R. Germany

Chapter 7

Psychological Research in Probabilistic Understanding

R.W. Scholz

This chapter is devoted to the research on probability which may be found in the subject of psychology, and deals with various research paradigms. Salient experimental tasks and research issues on how individuals cope with probabilistic settings are discussed. The objective is to provide a substantiated and representative review of the large number of psychological investigations. We start by presenting an indicative sample of psychological studies on people's response to probabilistic problems. Critical dimensions for judging the educational relevance of paradigms and issues will be introduced. The few developmental theories which deal with the acquisition of probability in psychology will be discussed. Shortcomings and perspectives of the educational research are critically examined in the concluding sections.

Research issues on probability occur among the subdisciplines of psychology. One finds important studies and ideas on how individuals react to probabilistic settings in such diverse areas as Cognitive Psychology, Social Psychology, Applied Psychology, Decision Research (in particular Risk Analysis), Mathematical Psychology, Developmental Psychology and Educational Psychology. The initial sections of this chapter have headings that may also be found in mathematical textbooks while the later sections are organized around psychological aspects.

For a long time most of the psychological research on learning, thinking, and teaching processes related to mathematics was dominated by an approach, in which the task and the subject were isolated and atomized. From an educational point of view, both a systems and a dynamic conception is necessary. So we discuss and critically appraise the conception of the task, of the subject, and in particular of the subject-task relation that underlies the reviewed psychological research.

The various paradigms are linked to different conceptions of man or rationality and to entirely different views on the subject-task relations. When discussing these different ap-

R. Kapadia and M. Borovcnik (eds.), Chance Encounters: Probability in Education, 213–254.

proaches the need to have a closer look at the whole scope of psychological probability research in order to understand and to illuminate stochastic thinking becomes clear.

1. Traditional Research Paradigms

Probability learning

The tasks. Consider the following paradigm: There are two light bulbs, left and right, and the subject has to predict which bulb will light up. The subject will be reinforced if the chosen bulb actually lights up. Let us assume that the probability, by which the two lamps light up, is given by a Bernoulli series, and let H denote the higher probability event, L the lower probability event.

The term probability learning was originated by Brunswik and Herma (1951). It refers to the finding that in the above paradigm the proportion of the subjects' choices tends to match the proportions of events (e.g., lamp lighting). From a psychological perspective, the paradigm under consideration is conceived as a learning situation with partial reinforcement. Many experimental studies on man and animals since 1930 were of this type, although the term probability learning was not applied at that time.

Probability learning experiments use a variety of experimental designs, among which the situation just described is the simplest. The number of stimuli was varied; extrinsic reinforcements were given by monetary payoffs; the label of the task was altered (problem solving vs. gambling experiment); the type of physical stimuli (lights, words, binary symbols or others) was changed. In addition, the structure of the response sequences was varied. Besides the Bernoulli series, Markov chains of order one were used to model the stimuli. In these Markov chains, the probabilities applicable on trial n depends on the outcome of trial n-1. Another variation consists of relating the reinforcement probability to the actual response given by the subject (contingent probability learning). In its simplest case of two outcomes there are two probabilities, the higher probability $P_h(H)$ on trial n if the subject chooses H on trial n-1 and $P_l(H)$ if the subject has chosen L on trial n-1. Further variants like probabilistic discrimination learning are described in Lee (1971).

Subject behaviour. Generally subjects started with a response frequency for the higher probability event H of about 0.5 and then adjusted towards the true stimuli probability $P(H)$ and beyond, before returning towards the response probability without reaching it.

Usually, only group average data were reported. For instance, Estes (1964) rejected attempts to analyze individual data or to draw conclusions about individual learning, because chance is considered to be a major variable in these individual results. Thus, the object of research was the prediction of group average data, which was often accomplished with high accuracy. However, if individual data are considered, the curves suggest considerable differences.

> "Certain subjects seem to adopt a strategy resulting from the frequency of reinforcements, which they later drop in favour of random response. Others adopt a maximizing strategy, in which only the most frequently reinforced response is used." (Fischbein, 1975, p.28)

One question, which has been widely discussed, is the rationality and appropriateness of the mean individual's choice. From a decision theoretical point of view, the maximizing strategy in the standard probability learning situation would be always to choose *H*, whereas probability matching (i.e. a strategy which randomly chooses *H* with P(*H*) and *L* with P(L) at each trial) yields an average hit rate $P(H)\,P(H) + P(L)\,P(L)$ which is inferior.

Theories of subject behaviour. Probability learning has been one of the favourite paradigms of behaviouristic psychological research. In line with the black box model of the cognitive system, the individual is perceived as being governed by random processes. Brunswik (1955, p.210) named this approach 'statistical behaviouristics'. Thus, probabilistic learning theories, like Estes' stimulus sampling theory or operator models (cf. Bush and Mosteller, 1955) were applied to describe the subjects' behaviour.

Let us consider a simple, but typical, application of the Bush and Mosteller model for the two alternative Bernoulli series. If the probability for the response *H* on trial *n* is denoted as p_n and *H* is reinforced on this trial, according to the linear operator model, the response probability on trial *n*+1 may be described by

$$p_{n+1} = \alpha_h p_n + (1-\alpha_h)\,\beta_h$$

where α_h and β_h are parameters independent of p_n. For the situation under consideration, β_h is the asymptotic response probability whereas α_h provides the relative increment or change rate. Usually, p_1 is assumed to equal 0.5. The parameters, α_h and β_h (if not assumed to equal the probability P(*H*) of the stimuli) are estimated by the data. If *L* occurs, the probabilities are adjusted by another operator equation with parameters α_l and β_l.

Such a poor theory for the conception of the human individual is worthless from an educational or developmental point of view. When, without special instruction, do the probabilities of the responses approximate the probabilities of the events? Is there a probability intuition like a geometrical intuition? To what extent can we speak of true stochastic conditioning, causing responses to match the probabilities of stimuli?

We should mention that some side effects of the early research on probability learning influenced current psychological theories; Jarvik (1951) observed that after a run of several identical outcomes, say *H*, subjects' response probabilities for this alternative 'fallaciously' dropped. This tendency to predict the outcome, which has not appeared for some time, was called 'negative recency' or 'gambler's fallacy'. Yet other typical patterns have been identified in the subjects' response series. Restle's (1961) so-called schema theory consists of two central ideas. The one is that the subject stores and pays attention to runs of the same event, the other, that when preparing a response the subject searches in memory until (s)he finds a run, which is at least as long as the current one. If the run, say of *L*'s is longer than the current one, *L* is predicted, otherwise *H*.

It has also repeatedly been doubted whether subjects actually consider the task to be random (Flood, 1954); this is, because the random mechanism is not usually obvious or subjects are not even told that they have to cope with a random sequence. Subjects

> "seem unable to recognize the simplicity of a Bernoulli series" and "therefore attempt to figure out the patterning involved." (Lee, 1971)

Bayesian revision

The task. The question of how well people use the information from data to update the probability that a hypothesis is true was much researched in the 1960s and is still frequently investigated (Slovic et al., 1986).

The following experiment illustrates the paradigm used in Bayesian revision studies. Subjects see two bags, one containing 70 red and 30 blue poker chips, the other containing 30 red and 70 blue; the first bag is called the red bag, the second, the blue. The experimenter flips a coin to choose one of the bags. He then begins to draw chips from the chosen bag. After each chip is drawn, the subject assesses the probability that one of the two bags, e.g., the red one, is chosen. According to the Bayesian approach the normative solution is provided by Bayes' rule

$$P(H_i|D) = \frac{P(D|H_i)P(H_i)}{\sum_i P(D|H_i)P(H_i)}$$

where $P(H_i|D)$ is the posterior probability that a certain hypothesis H_i is true, taking into account a datum D. For instance, for the hypothesis that the chips were drawn from the red bag and the data that 8 reds and 4 blues were chosen, a posterior probability of 0.97 for the red bag can be calculated. As usual, many task parameters including the nature of the probability apparatus were varied (cf. Lee, 1971).

Subject behaviour. Subjects' behaviour was compared with the Bayesian solution, which was considered the optimal one. In the above bag example, most subjects gave an assessment between 0.7 and 0.8. as opposed to 0.97. Rapoport and Wallsten (1972) report that in most studies

> "the estimates were monotonously related to these quantities". (p.153)

Furthermore, subjects updated their probability estimation to a much smaller extent than Bayes' rule prescribes. This phenomenon is called conservatism.

Theories of subject behaviour. In these studies man is considered as an automaton who revises his/her probability according to Bayes' rule. For a long time, theories almost exclusively focused on explanations of the deviating behaviour from the optimal response, i.e, the conservatism effect. Models were usually modifications of Bayes' theorem by introducing weights to decrease the impact of the diagnostic information. Conservatism was even regarded as an example of the so-called central tendency effect of the well-known adaption theory (cf. Helson, 1964), which was originally designed to model changes in perception with changes of stimuli, for instance the widening of the diameter of pupils with the changes in intensity of illumination.

Slovic et al. (1986) claim empirical support that the subjects' conservatism is attributable to

> "(a) misunderstanding of the data-generating process and, thus, of the diagnostic impact of data, $P(D|H_i)$; (b) inability to aggregate the information received; (c) unwillingness to 'use up' the bounded response scale, knowing that more data were forthcoming." (p.682)

Later explanations rejected Bayes' rule as a good descriptive model of subjects' behaviour. Kahneman and Tversky (1972) summarize:

> "In his evaluation of evidence, man is apparently not a conservative Bayesian: he is not Bayesian at all." (p.450)

Disjunctive and conjunctive probabilities

The tasks. Studies conducted on individuals' responses to disjunctive and conjunctive probabilities have been conceptually restricted to tests of the behavioural validity of the probability calculus. One of the most elementary rules is the addition theorem. The question of whether probability estimates of a set of mutually exclusive and exhaustive events sum up to 1, was investigated in one of the classical studies of Cohen et al. (1956). In a frequentist setting children aged 9-14 had to estimate the 'subjective probability' of success and failure in a series of ball throws through differently sized target apertures.

Another elementary probability rule is the so-called extension rule, i.e., if events A and B are such that $A \subset B$, then $P(A) \leq P(B)$. In a series of recent experiments on subjects' ability to assess the probability of conjunctive events, experimental psychologists used simple word problems. The subjects are asked to compare the likelihood of the events A and B where A means a massive flood somewhere in North America in 1989, in which more than 1000 people drown, and B means an earthquake in California sometime in 1989, causing a flood in which more than 1000 people drown. Clearly, the event B is a subset of A as California is part of North America and not all floods are caused by earthquakes.

Most studies of the subjective probabilities of conjunctions, which focused on testing the multiplicative rule $P(A \wedge B) = P(B)\,P(A|B)$, only considered the special case of independent events, though, the abstract problem structure has been put in different settings. For instance, Scholz and Köntopp (1990) compare subject responses in classical two urn problems to those in tasks with an everyday setting like the probability of reaching a certain bus and getting a seat.

Subject behaviour. In the analysis of the above mentioned ball-throwing experiment, Cohen et al. added the number of successes and failures reported by different subject samples. Since the samples were comparable with respect to age, social status, and educational level, the sum of the 'subjective probabilities' (i.e. relative frequency of estimates) of success and failure should approximately add up to one. However, this was obviously only valid for very easy tasks (in which the success was certain) or extremely hard ones. Furthermore, an effect of the response scale was found. In some of the studies subjects had to report their estimates of the probability by the absolute number of successes or failures. If, for instance, five trials were considered as the reference 'scale', the sum of the subjects' estimates of probabilities of failures and successes was about one; if many trials (100 or 1000) are referred to for responding, the sum is considerably below one.

With the 'flood problem', more than 200 undergraduate students first had to evaluate the probability of several events on a nine-point scale, 0.01%, 0.1%. 0.5%, 1%, 2%, 5%, 19%, 25%, and 50%. Half of the subjects evaluated the probability of the event A, and the other half B.

> "The estimates of the conjunction (earthquake and flood) were significantly higher than the estimates of the flood ... The respective geometric means were 3.1% and 2.2%. Thus, a suggestion that a devastating flood could be caused by the anticipated California earthquake made the conjunction of an earthquake and a flood appear more probable than a flood. The same pattern was observed in other problems." (Tversky and Kahneman, 1983, p.307)

Usually, studies on the multiplicative rule conclude with the statement that people generally overestimate the probability of conjunctions or state roughly that the mean response on conjunctive problems is somewhere between the product and the mean of the probability of the events. When categorizing typical response patterns, Ross and de Groot (1982) reported that a main error was equating the probability of a proper intersection with that of one of its components, or indeed arithmetically averaging the probabilities.

Theories of subject behaviour. Violations of the extension rule are labelled as the *conjunction fallacy*. Psychologists again and again revealed serious violations of even the most elementary rules of probability. How are these deviations explained? What theories are offered for this phenomenon?

In the early research, the 'behavioural' validity of probability rules was not immediately abandoned. Cohen et al. (1956) believed that the nonadditivity is attributable to numerical difficulties; subjects are familiar with specific numbers or ranges and have difficulties in using different ranges. This type of reasoning may be regarded as a typical scale artefact argument.

Nowadays one would say that most explanations for individuals' behaviour were just ad hoc hypotheses of the scale artefact type. Falk and McGregor (1983) revealed that in judgements on the probability of rare events it obviously matters whether or not one is involved in the situation under consideration. They argue that a person might

> "perceive himself as one in a set of one, while you .... are perceived as one in a set of many". (p.489)

Biases linked to egocentrism were also convincingly demonstrated by Svenson (1981); the average individual is prone to consider her/himself to be more skilled (as a driver, for example) than the average individual. Another common argument used for the explanation of the effects are differential factors like gender.

In a recent, but traditionally orientated approach, Hogarth (1980) explained the subjects' responses in conjunctive tasks through a simple information weight model. Let C denote a potential cause or contributing factor and E a presumed effect/event. According to Hogarth the judged probability of the conjunction may be written as

$$P(C \cap E) = [P(E)\,P(C)]^{a} \quad \text{with} \quad a = 1 - P(E|C)\,.$$

Hogarth states that this model provides a good account of judgements in a variety of tasks. It reflects meaningful dependencies on task parameters; the conjunctive probability increases with $P(E)$, $P(C)$, and $P(E|C)$. Thus the model provides a simple, general, and predictive plausible model of the phenomenon of the *conjunction fallacy*; it also allows for reference to the framework of inferential heuristics.

We would like to remind the reader that several critical issues and questions were not dealt with in this section. For instance, one has to ask which notion of probability is needed in a specific problem and whether it is at the subject's disposal. One should also note that most psychologists refer to the notion of subjective probability, while forgetting that subjectivists consider independence a derived concept. The lesson which may be learnt from this section is that these experiments show that there is no coherence in probability judgements, whether deliberate or not.

Correlation

The tasks. There are two features of the experimental tasks considered now which differ from those reported in the 'probability learning' section. First, subjects are asked to express their degree of perceived correlation between two events (usually by a number between 1 and 100) instead of predicting the occurrence of future event combinations. Second, most studies use contexts from the subject's environment; context free tasks such as 2x2 contingency tables with general labels (psychophysical tasks) are used for control purposes. Most early research was conducted on people's ability to estimate associations correctly from fourfold, presence-absence tables.

The rows and columns had labels like the frequency of wins of a soccer team for phenomenon I and the weather condition (warm vs. not warm) for phenomenon II. In one provocative series of experiments, the psychologists' practice itself was made the object

Phenomenon I	Phenomenon II Present	Absent
Present	100 Cell *A*	50 Cell *B*
Absent	400 Cell *C*	200 Cell *D*

Table 1: Presence-absence table for two variables

of study; Chapman and Chapman (1967) investigated the validity of various accepted clinical tests like the Draw-a-Person test. Many clinicians believe that this test is a valid diagnostic instrument and that paranoid people like hyper-suspicious ones more frequently produce pictures with big eyes than others. After practising clinicians had studied the covariation of symptoms and features of subjects, diagnoses of drawings were randomly presented to (relatively naive) psychology students. The students were then asked which sign is most frequently associated with a symptom (e.g., suspiciousness).

Arkes and Harkness (1983) used a series of experiments with 2x2 tables to investigate how subjects' assessment of contingency is guided in such tasks. After controlling the effect of similar column and row labelling, they systematically varied the cell frequencies in order to find out how many subjects base their contingency judgement solely on the frequency of Cell A, on the difference between the frequencies of Cell *A* and B, the sum of Cell *A* and D, or the difference between the sum of the diagonals.

Subject behaviour. Although, in the Chapmans' experiment no correlation between symptoms and drawings should have been reported by the naive students, they seemed to have seen the same covariations in random data that the clinicians had claimed to see in their practice. There are also conclusive findings from the Arkes and Harkness' study

- the entry of Cell *A* appears to have a major impact on contingency estimates
- labelling and arrangement of tables can drastically influence the contingency judgements
- low cell frequencies result in overestimation of correlation
- subjects consistently use a particular heuristic (e.g. the *A-B* heuristic)
- less complex heuristics are applied when the memory demands are increased.

Theories of subject behaviour. The Chapmans' study reveals an essential misconception. Obviously, there are common expectations and beliefs about the relationship between events that cause the impression of empirical contingencies. Furthermore, these beliefs are maintained in spite of evidence of the independence of events. These phenomena are subsumed under the label *illusionary correlation*. The story of the illusionary correlation studies demonstrates that one of the fundamental assumptions underlying most of the research reported in earlier sections is highly doubtful. The notion of covariation is by no means theory-free, context-free and automaton-like; rather, the information processing should be interpreted as theory-driven.

Let us consider the lessons to be drawn from more recent research on covariations in 2x2 contingency tables. First, the exclusive reliance upon the 'Present/Present' cell seems to be a particular failure that is increased when all cell frequencies are not presented simultaneously. Second, other systematic subsidiary strategies like the 'sum of main diagonal heuristic' are applied; furthermore these strategies and the judgement of most subjects are susceptible to the display mode and the perceptual prominence of low numbers. Hence, Nisbett and Ross (1980) conclude

> "Without formal statistical training, very few people intuitively understand that no judgement of association can be made legitimately without simultaneously considering all four cells." (p.91)

The heuristic, which individuals apply, varies not only interpersonally, but also intrapersonally. Subjects show a multitude of cognitive strategies with the same problem structure; situational and framing factors also influence the heuristic that is applied. Arkes and Harkness (1983) make the following radical observation.

> "All of these results lead us to the conclusion that a search for the heuristic that people use will be futile. The human information processing system is flexible enough to shift strategies depending on task characteristics." (p.132)

2. Current Research Paradigms

Judgemental heuristics

The tasks. It has been shown that human judgement and information processing is guided by a person's expectations, beliefs, and experience. Of course, this implies a significant shift from the behaviourally oriented 'automaton' conception of human action which under-pinned earlier research. Judgemental heuristics provide a somewhat different approach; in their experimental studies Kahneman and Tversky (1973) tried to prove that individuals' probability judgements are systematically biased by the use of few inferential heuristics.

In the 'almanac problem' the frequency of letters in the English language was studied. The instructions reads as follows. Consider the letter *R*. Is *R* more likely to appear in the first position? the third position? My estimates for the ratio of these two values is ___:1.
The 'correct' response for this problem is based on extensive word-counts; altogether eight consonants appear more frequently in the third than in the first position of which (*K,L,N,R,V*) were selected for investigation.

The 'base-rate problem' reads as follows. A cab was involved in a hit-and-run accident at night. Two cab companies, the Green and the Blue, operate in the city. You are given the following data: (i) 85% of the cabs in the city are Green and 15% are Blue. (ii) A witness identified the cab as Blue. The court tested his ability to identify cabs under the appropriate visibility conditions. When presented with a sample of cabs (half of which were Blue and half of which were Green), the witness made correct identifications in 80% of the cases and erred in 20% of the cases. What is the probability that the cab involved in the accident was Blue rather than Green?

In a 'base-rate problem', two sorts of information are given. There is the base-rate information which usually provides the starting point in the individual's information processing about a hypothesis H_1; this is the rate of Blues and Greens in the problem. It is often - but not always - provided by statistical data. There is also indicative or additional information; in the Cab problem, this is the reliability of the witness. Usually, this information or datum will be denoted by D. Often, this datum provides information if the hypothesis H_1 or its complement H_2 is operative; D represents diagnostic information. Responses on base-rate problems which coincide with the diagnostic information are called diagnostic responses.

The 'Cab problem' does have an accepted 'normative solution' which can be derived from Bayes' theorem. The requested probability for a Blue cab, given the witness' testimony D that the cab is Blue, will be denoted as $P(H_1|D)$. In the problem, the base-rates are $P(H_1)=0.15$ and $P(H_2)=0.85$ the diagnostic information yields a 80% chance for a Blue testimony given the cab is Blue, $P(D|H_1)=0.8$ and a 20% chance for a Blue testimony in case of a Green, $P(D|H_2)=0.2$). From Bayes' theorem one gets $P(H_1|D)=0.41$.

Actually, there are various degrees of freedom in task analysis in this problem which have been elaborated by Scholz (1987). Mapping between words and mathematics is very delicate; a critical aspect of written problem tasks is the incompleteness or indeterminacy of the description of a real setting. The cues and parameters of real life problems are unlimited so that it is rare to have a unique solution. In order to obtain such a unique solution, restriction to a few important features is necessary. However, confronted with the 'Cab problem', a subject may introduce more or less sophisticated assumptions about the activity rates of the Blue and the Green companies. These assumptions may lead to modifications of base-rates and produce a different yet reasonable answer even within the Bayesian calculus.

Scholz (1987) sees adding compatible new information as a special case of a phenomenon called introducing a theory about the information; there are several possibilities for such reasonable modifications. The normative solution assumes that the identification rate is independent of the actual cab colours. However, there is strong evidence from signal detection experiments that the hit rate depends on the colours (cf. Green and Swets, 1966). Thus the 80% hit rate assessed with half Blues and Greens, may not be appropriate for the actual base-rates of 15% and 85%. The independence of the diagnosticity from the base-rate is unrealistic in many other base-rate examples.

Moreover, the parameters of base-rate problems may be transformed by reassessing the values according to different conceptions of probability. In accordance with subjectivist probability a subject may assume that the credibility of a witness does not always exactly equal 0.8. If a symmetric probability distribution around the 0.8 value is assumed, then the solution differs from the 'normative' one. It can be shown mathematically that if the expected value is kept constant, a greater variance in the diagnostic distribution yields a higher posterior probability in the problem. Therefore, base rate problems do not have a unique solution.

Another base rate problem is the so-called 'Tom W. paradigm'. Subjects were given a brief personality description. They were told that this description has been written by a panel of experts and then randomly drawn from a pile of descriptions of say 70 engineers and 30 lawyers. The experimental task was to estimate the probability of Tom W. being an engineer. A typical text was: Tom W. is a 45-year-old man. He is married and has four children. He is generally conservative, careful, and ambitious. He shows no interest in political and social issues and spends most of his free time on his many hobbies, which include home carpentry, sailing and mathematical puzzles.

Subject behaviour. Among the 152 subjects in the almanac problem, 105 judged the first position to be more likely and 47 judged the third position to be more likely for a majority of the letters. Moreover, each of the five letters was judged by a majority of subjects to be more frequent in the first than in the third position. The median estimated ratio was 2:1 for each of the five letters (Tversky and Kahneman, 1979, p.211).

Experimental studies analyzed subjects' neglect of base-rate information, when they are exposed to diagnostic information. This phenomenon is called *base-rate fallacy*. For a long time, the analysis focused almost exclusively on the percentage of subjects who responded with diagnosticity. It is surprising that researchers did not investigate those subjects not responding with diagnosticity. In the Cab problem, the modal and the median response was 80% which coincides exactly with the credibility of the witness and ignores the base-rates. In fact, the percentage of diagnostic responses did considerably vary between the studies from 13% up to 50%.

However, different factors like context which affected the variability of response distributions were investigated. Tversky and Kahneman (1973) compare two groups of college students in the Tom W. paradigm. One group were told that the descriptions were drawn from a file of 70 engineers and 30 lawyers, whereas the other group were told that the pile had 30 engineers and 70 lawyers. Subjects' estimates were only slightly (though statistically significantly) affected by the base-rate information.

Theories of subject behaviour. According to Tversky and Kahneman, probability judgements are systematically biased, because people often apply judgemental heuristics. The most important of these are the availability heuristic, the representativeness heuristic, the causal schema, and the specificity construct. The representativeness heuristic is the most extensively discussed.

> "A person who follows this heuristic evaluates the probability of an uncertain event, or a sample, by the degree to which it is (i) similar in essential properties to its parent population; and (ii) reflects the salient features of the process by which it is generated." (Kahneman and Tversky, 1972, p.431)

In the Tom W. paradigm, subjects' answers may be due to the representativeness heuristic as people tend to judge the likelihood of Tom W. being an engineer by the extent to which the description of the person is similar to an engineer; they do not consider the statistical information. Representativeness may also be considered to underlie the gambler's fallacy. If a gambler is involved in a series of Heads and Tails with a fair coin and has experienced a series *TTHHH*, he is more prone to expect a *T* in a forthcoming sixth trial than a *H*, because he considers a series with an equal number of Heads and Tails to be more representative than a series of two Tails and four Heads.

In the 'word frequency problem', availability is thought to cause the response distributions, as it is easier to think of words that start with *K* than of words where K is in the third position.

> "A person is said to employ the availability heuristic whenever he estimates frequency or probability by the ease with which instances or associations come to mind." (Tversky and Kahneman, 1973, p.208).

Originally, the *base-rate fallacy* was believed to be a pervasive shortcoming of human performance. As experiments have shown that base-rates are sometimes considered and sometimes ignored, the availability heuristic was used to explain the neglect of base-rates. They tend to be ignored, if they are remote and abstract, or if the numbers in the problems are hard to recall or to use in calculation. On the other hand, base-rates are taken into account, if they are vivid, salient, and concrete. The diagnosticity responses in the Cab problem may be caused by the base-rates being perceived as abstract (Nisbett and Ross, 1980).

The specificity construct may also explain the *base-rate fallacy* (Bar-Hillel, 1980), according to which information is more specific and dominates other information, if it refers to a smaller subset. In the case of the Cab problem, for instance, the diagnosticity information refers to just the one taxi and hence is more specific than the base-rate information, whereas in other problems both the base-rate and the diagnosticity information may refer to the same issue and are thus equally specific.

According to the causality construct, the following holds. If the base-rate information is causally linked with the event, for which a probability judgement is required, it is taken into account. In particular, in the Cab problem, the base-rates are causally linked if two companies of equal size are considered to produce different accident rates. The subjects' responses should be near to the 'normative solution' if the constituents of the base-rate are causally linked to the probability of an accident whereas there is no such causality link in the original version of the Cab problem.

The heuristic approach entails several deficiencies. For instance, heuristics in the field of decision and judgement are fuzzily defined and show few characteristics of the kind used in cognitive psychology, problem solving, or artificial intelligence. Although, judgemental heuristics clearly embody a major step forward toward an understanding of cognitive processes in stochastic thinking, they have been criticized for methodological and conceptual reasons. Hardly any knowledge had been acquired about comprehension of the text by subjects, how the situation is perceived, the meaning given to concepts and the tools available. Clearly an answer to these questions implies a change in applied research methodology. From a theoretical perspective, a severe criticism is that the different heuristics and microprocesses are defined in isolation from each other, side by side, without any theoretical considerations about their mutual interdependence. There is no general model of cognitive activity, nor are there rules determining which heuristic is selected.

Structure and process models of thinking

Clearly, the critique of paradigms discussed requires a change in methodology. In what follows an approach by the author is summarized. Models of the structure and process of thinking are developed to specify the different types of knowledge and inferences involved in problem solving and stochastic thinking (see Scholz, 1987).

The tasks. For an analysis of human inference and cognition, the subject's understanding and internal representation of the text, his/her goals, and the individual's information processing and conclusions, are indispensable objects of study. If one wants to study whether information is causally interpreted, whether probabilities are transformed or weighted, or whether probability judgements are actually based on representativeness considerations, one must design the experimental procedure to produce data which allow for a tracing of the subjects' thought processes.

A special procedure was designed to carefully study subjects' understanding of text and question, and to follow their thought processes. Subjects were introduced to a task by listening to a tape-recorded text and then read it on a word processor at their own speed making notes if necessary. After they had finished reading the text and questions, subjects had to reproduce and write down the text and the questions on another sheet. Not until then were they asked to tackle the problem. They were asked to provide an extensive protocol on the thought process, reasoning and associations, which led them to an answer. To make sure that the subjects did engage with the experiment, they were (unexpectedly) asked after the first participation to repeat this painstaking procedure, yet a third time at home.

The conception of a stochastic problem does not only depend on the underlying mathematical model. Generalizing Bar-Hillel's (1980) distinction between a textbook and a social judgement paradigm, we introduced the constructs of a problem solving frame and a social judgement frame of a task, specifying typical features of both the content and the circumstances in which a task is presented to an individual. This distinction is essential for an interpretation of the learner's activity.

In a *problem solving frame*, tasks have a codified form with a mathematical setting, where information has to be taken for granted. Though contexts come from the technical world, the information appears to be precisely and objectively assessed. There are usually a few, sequentially introduced cues embedded in a closed story, scarcely permitting the introduction of additional information. Consequently, one can usually agree upon a single, unique solution.

In a *social judgement frame*, the content appears to be realistic and taken from the non-technical world. The information is not completely precise; there are a large number of redundant cues, which may be linked to different personal experience and hence result in different solutions. Individuals are not expected to produce an exact answer, but rather an estimate based on personal expertise and experience.

Subject behaviour. A wide range of subjects were investigated, including 7th graders, college students and ten university professors, who specialized in stochastics. Subjects were able to reproduce the text of the problem literally. However, a considerable proportion of them started from other interpretations of the question than the one intended by the researcher. A common error was to start with the inverse probability. This means that instead of $P(H|D)$, they thought about $P(D|H)$, which is the 'diagnosticity response', so that the label 'base-rate fallacy' seems unjustified.

The protocols on the thought process reveal that a multitude of strategies were applied by the subjects, yet indications for a use of the heuristics above could only be found occasionally. An exploratory analysis of more than two hundred protocols suggested that two different modes of thought guide the judgemental process. A complementarity of intuitive and analytic thinking was introduced by lists of features, which were assigned by raters to the written protocols of a subject's thought process. Various studies by four independent raters have shown that these modes can be reliably discriminated.

The protocol analysis also indicates that the knowledge and the heuristics depend both on subjects' background (age, educational level etc.) and on the way a problem is framed. Natural science students looked for a precise algebraic solution, either the normative or diagnosticity one, avoiding fuzzy middling responses or rough probability judgements, whereas analytic-algebraic strategies and the diagnosticity answers were only rarely found in a social science sample.

There are obviously forms of task presentation which enhance a formal, analytic treatment of the problem and others which elicit rough-and-ready heuristics. 'Normative solutions' are not necessarily beyond students' capability, one third arrived at them. Another finding of theoretical importance was that diagnosticity responses were not primarily due to the above 'intuitive heuristics' but to 'analytic' reasoning.

Theories of subject behaviour. The results have led to a framework for the structure and process of thinking in information processing (SPT models), which allow for a more differentiated understanding of stochastic thinking and provide also an approach for integrating the various rather atomistically formulated judgemental heuristics. The first step in this approach is structuring the multitude of strategies by the complementarity of intuitive and analytic thinking. A description of these modes of cognitive activity is given in the table below.

	Intuitive thought	**analytic thought**
A	preconscious - information acquisition - processing of information	conscious - information acquisition and selection - processing of information
B	understanding by feeling, instinct of empathy	pure logical reasoning, independent of temporary moods and physiology
C	sudden, synthetical, parallel processing of knowledge	sequential, linear step-by-step ordered cognitive activity
D	treating the problem structure as a whole, 'Gestalt erkennend'	separating details of information
E	dependent on personal experience	independent of personal experience
F	pictorial metaphors	conceptual or numerical patterns
G	low cognitive control	high cognitive control
H	emotional involvement without anxiety	cold, emotion-free activity
I	feeling of certainty toward the product of thinking	uncertainty toward the product of thinking

Table 2: Attributes and features of intuitive and analytic thought (Scholz, 1987, p.62)

These 'modes of thought' influence probability judgements and allow insights into the nature of information processing and fallacious responses. A finding which contradicts judgemental heuristic research, is that the diagnosticity responses in base-rate problems were not rated to be a product of intuitive heuristics but were judged to result from analytic thought processes. Some students are prone to produce crude and unsophisticated responses, mainly in the analytic mode; often the calculations performed seem to be the product of a blind, insightless, and mechanistic process, as if a conditioned response pattern is elicited, roughly taking the form: probability - mathematics - pick up the numbers - produce a calculation - perhaps it fits.

The framework is based on a systems theoretical approach rather than a computer analogy; it is shown in figure 1. The nine units of the framework each have a different status. The main processing units are the Working Memory including a Short-Term Memory and the Guiding System. The four units of Long-Term Storage are the Knowledge Base, the Heuristic Structure, the Goal System, and the Evaluative Structure; these may be ordered along two dimensions. First, the Heuristic and Evaluative Structures are assumed to consist of operators which may be applied to the entities of the Knowledge Base, the Goal System, and of course encoded information. Second, we distinguish stores used in the direct dealing with the problem under consideration (i.e., the Knowledge Base and the Heuristic Structure) from stores, which are involved in the evaluation and direction of the cognitive system (i.e., the Evaluative Structure and the Goal System).

There are various reasons for these distinctions. For instance, the acquisition of new operations takes more time than the learning of new concepts or new goals. Further, it is a common feature in clinical psychology that some person's Knowledge Base and Heuristic Structure are highly developed, yet severe deficiencies may be identified in goal setting behaviour and self-evaluation. In general, we presume that in both dimensions a different length of time is required for the acquisition of new elements. On the one hand, new knowledge elements and new goals are usually acquired more quickly than heuristics or evaluative operators. On the other hand, goals and evaluators are acquired more slowly than the corresponding operative entities (i.e., elements of the Knowledge Base of Heuristic Structure).

Finally, within our cognitive framework there remain the Sensory System, the Decision Filter, and the (overt) Action, which may be regarded as both part of the individual and his/her environment.

We believe that the proposed framework for information processing in stochastic thinking provides a basis for an integration of classical heuristics. The availability heuristic is primarily based on processes, which are localized in the initial phase of information processing, i.e., the encoding and retrieval, although availability may be considered as a very general concept, which may concern each activation process, for instance, any activation of heuristics, goals, or evaluative operations. The representativeness heuristic, as a global similarity matching heuristic, has to be regarded as a typical, simple everyday heuristic. However, representativeness may also be determined by analytic levels.

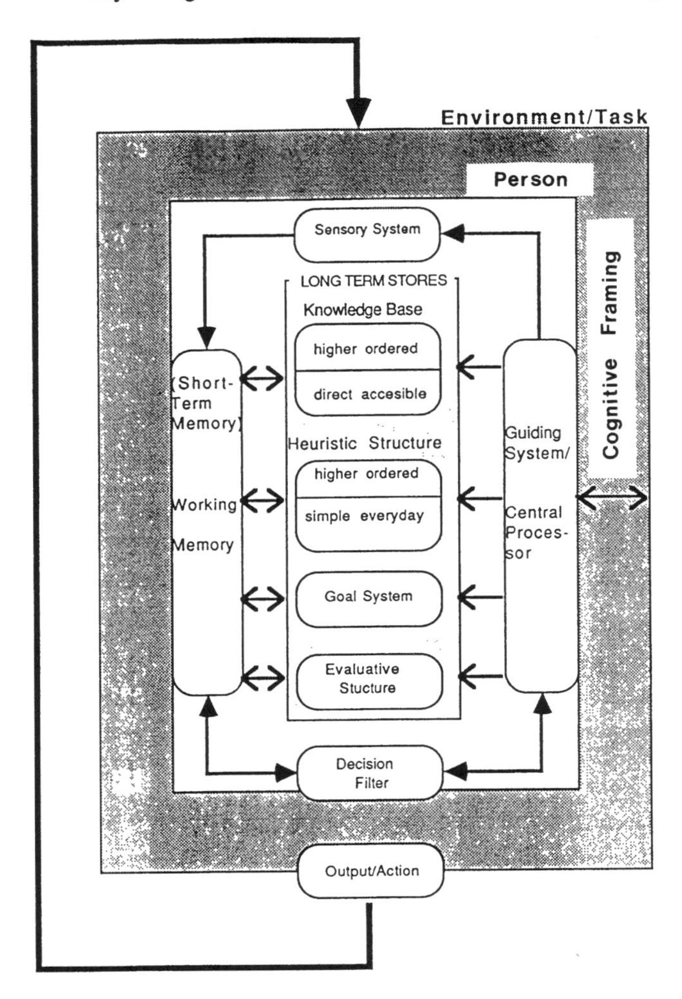

Fig.1: A framework of the structure of information processing: the general model

Within the framework, the causal schema has to be conceived as a more complex cognitive activity that not only involves the Heuristic Structure but presumably also the Evaluative Structure. The causality between entities is seldom present in the Knowledge Base, it is nearly always the result of an evaluative and comparative procedure. This reveals that heuristics encompass essentially different processes which are better separated conceptually.

Within the protocol analyses further heuristics have been identified such as rough choice strategies between displayed numerical values or intuitive information weighing. Mathematical theorems, probability concepts, tests or inferences learnt at school may also be embedded in the higher ordered heuristics. However, these entities are two-sided. They possess a static, conceptual, knowledge-based side, and also an operative, active, dynamic heuristic side. Especially for stochastic notions, a connection may be hypothesized between the static conceptual elements that are activated and the heuristics that are applied.

In order to avoid misunderstandings on the modes of thought concept we will discuss some of their features. First, the intuitive mode is not conceptualized as being the same as an analytic mode, on a simpler and lower level, but rather as an essentially different activity in which other domains of long-term stores are involved. In general, no behavioural superiority of one of the two modes is to be expected in stochastic thinking.

Second, each of the modes of thought specifies the quality and intensity of actual information processing. We believe the intuitive mode to be the way of thinking that is most naturally chosen unless an analytic mode is induced by a switch through the Central Processor. In the intuitive mode, information is sampled in a relatively haphazard manner. Within the Knowledge Base, the directly accessible knowledge is retrieved and used in the encoding process; there is no systematic search in either the Knowledge Base or the Heuristic Structure, and mostly simple everyday heuristics are applied.

Third, in analytic thinking more units are involved; the domains of higher ordered knowledge and heuristics are pre-initialized (top-down), and the Evaluative Structure is activated so that evaluative operators are fed with high density to the Working Memory. The information acquisition itself is accompanied by a guided selective information search and a systematic focusing exploration in the Knowledge Base (bottom-up) within the pre-initialized domain. Sequential, higher ordered, and sometimes even formal heuristics are made available, and the information processing is accompanied by a self-reflective, self-controlling monitoring. The decisions are controlled by the Decision Filter, and fed back to the Working Memory and the Central Processor.

Probability calibration

Behavioural Decision Theory is becoming an elaborate and organized field which aims at the description of judgement, inference, and decision making techniques. Probability is only one constituent of decision making; nevertheless a considerable part of research on probability judgement was carried out by decision researchers and decision analysts. Based on a model of structuring the decision process, decision analysts usually separately measure (subjective) utility and probabilities. Later, these entities are (mechanically) combined for instance by linear models or within the framework of decision trees (cf. Winterfeldt and Edwards, 1986). A review of the scope of methods and software tools for structuring and analyzing decision problems is in Humphreys and Wishuda (1987).

As probability is one constituent of such procedures they require subroutines for probability assessment. In order to get the right values, a variety of direct or indirect techniques were developed. Professional decision analysts frequently use spinner tasks in order to measure probabilities for binary events as demonstrated in the following example.

The tasks. To assess the probabilities of event E, the wheel is adjusted until the judge is indifferent between two bets. In the context of weather forecasting an example may read: You win ten dollars if it rains tomorrow, nothing if it does not rain (Gamble 1). You win 10 dollars, if a spinner, spun on a wheel, lands in the black sector which covers 80% of the circle (Gamble 2).

The most widespread method for judging probability ratings is the calibration paradigm. A widely used technique of typical questionnaire studies uses the so-called half range method of calibration. Subjects must answer a question and then assign a subjective probability on their certainty about their answer. For example:

Which city is further north? a) New York b) Rome
Subjective probability 50---60---70---80---90---100 %

Subject behaviour. May (1986) distinguishes between two types of calibration studies. The traditional approach consists of comparing calibration under different sets of tasks; items are categorized by the respective subjective probability. Using, for instance, the above half-range method, the subjective probabilities are compared with the percentage of correct responses. The reviews by Lichtenstein et al. (1982, p.314) point out:

> "The most pervasive finding in recent research is that people are overconfident with general knowledge items of moderate or extreme difficulty."

Overconfidence increases toward the extremes of the probability continuum and was generally considered to be independent of the response format or the specific method (i.e. half versus full range method). They report a study using odds in which answers with 1,000:1 odds were only 81% to 88% correct. However, studies in professional contexts yield much better performance: Murphy and Winkler (1974) found average deviations of only .028 from the perfect calibration for credible temperatures for weather forecasters.

Theories of subject behaviour. As noted above, decision researchers believed in the 'overconfidence' phenomenon for a long period and certainly a considerable number of researchers still believe in this robust effect; thus it is not surprising, that a series of explanations are offered. Pitz (1974) argued that in a series of inferences the uncertainty in the earlier stages may not be carried over in the later stages. Others believe in response scale artifacts or the general inferential shortcoming hypothesis. One popular variant is that people become overconfident as they search their memory only for confirming evidence, while neglecting contradictory evidence.

These explanations were criticized by May (1986) who did not analyze classes of items, but carefully constructed a specific cognitive model for each single item. In order to understand miscalibration she analyzed the factors, the subjective probability rating and the determinants of right and wrong answers. She concludes

> "The normative idea of calibration and the derogatory interpretation of 'overconfidence' as a general 'shortcoming' turned out to be critical: confidence based on wrong knowledge is 'overconfidence' by definition." (p.28)

In the light of this approach, the nearly perfect calibration of weather forecasters is not surprising. United States weather forecasters have been making public probabilistic forecasts since at least 1965. One may assume that in their natural context wrong knowledge is eliminated through the course of experience so their knowledge base is not distorted but well adjusted toward the object of their judgements.

Event-related brain potential research

Variants of probability learning and the concept of subjective probability play a major role in current controversies on event-related brain potentials. An event-related brain potential (ERP) is a transient sequence of voltage fluctuations induced in brain tissue by the occurrence of some critical event (cf. Horst et al., 1980). One of the standard paradigms is the so-called 'oddball paradigm'. These are tasks in which random series of two or more stimuli are presented, and subjects must count or respond to one of the stimuli or they must respond differently to either stimulus. When doing this job the subject wears EEG-electrodes (electroencephalogram) on his scalp by which the brain potential is recorded.

The critical measure in brain potential research on subjective probability is the so-called P300 which is a peaking on the EEG at about 300-600 msec after certain stimulus onset.

Horst et al. (1980) summarize the main finding of this branch with respect to expectancy which, for a long time, was synonymously used for subjective probability:

> "Our data indicate that the amplitude of the P300 ... was determined by the interaction between a trial's outcome and the subject's expectancy concerning that outcome ... These data ... strengthen the claim that P300 amplitude is dependent on the subjective probability associated with the ERP-eliciting event." (p.483-484)

We will end the discussion of the paradigms by an instructive dispute on subjective probability in recent P300 research. The (traditional) position states that a P300 is a manifestation of activity occurring whenever one's model of the environment must be revised (cf. Donchin and Coles, 1988). Thus in Bernoulli series a P300 appears if the rare event 'surprises' the subject's expectancy. This theory is called 'context update theory'. Another theory believes that subjects maintain an internal template of the context, which includes maintaining an expectancy of the event that will close the context. Thus when the stimulus is the expected one, stimulus evaluation leads to response selection on the one hand and to closure of the present context on the other. This theory is called 'context closure'.

Kahneman and Tversky (1982) discriminate between different 'variants of uncertainty'. The first type is an "active expectation which occupies consciousness and draws on the limited capacities of attention". The other two are types of automatic passive expectations. One of them, the permanent type, consists of expectations about covariations of attributes which are used to organize and encode experience. The other version comprises temporary expectations (priming). In contrast to active expectations, passive expectations do not impede the response to targets that have not been primed. They go on to relate their differentiation of variants of subjective probability to the P300 research:

> "... a run of repetitions of the same event is associated with a steadily decreasing P300, suggesting an increase in the subjective probability of further repetitions. In contrast, the conscious expectations of repetitions decreases consistently during a long run, by the familiar gambler's fallacy. Evidently, an observer can be prepared, or 'primed' for one event while consciously expecting another - and show physiological evidence of surprise at the occurrence of an event that was consciously predicted. Thus, there is a sense in which an individual can have conflicting probabilities for the same event at the same time. These observations suggest an image of the mind as a bureaucracy ... in which different parts have access to different data .." (p.512)

Karis et al. (1983, p.265) postulate a rather rigid version of passive subjective probability assessment based on automatic, subconscious, autonomous processes of expectancy formation.

> "There is, however, no need to assume that the assessment of subjective probability is in any way conscious ... The computations may be performed by internal processors designed to assess the probability structure of the environment."

If this version is valid and this type of probability processor is believed to be 'at work' in the gambler's fallacy, the P300 may be explained. Note that within the above, traditional 'surprise interpretation' of a P300, a contradiction arises, if the active conscious version of subjective probability is actually meant. In a Bernoulli series the subjective probability of the repeated event decreases. The rare event becomes the more likely, the longer it does not appear and, thus it is less and less surprising if it finally appears.

Verleger (1988, p.13) doubts whether it is legitimate to postulate a subconscious expectancy formation. He believes that subjects are conscious that their task is 'to wait for the rare stimuli', that is to wait for the closing stimuli of an epoch resulting in the active conscious version of expectancy; so a contradiction would rise within the brain potential approach. This controversy is highly significant for substantiating one of the core assumptions of the SPT-models, namely that subjective probability may be assessed in completely different manners. The scope of areas, fields and research paradigms in which probability research is conducted is broad. The discussion on the brain potential approach implicitly means that psychological probability research is theoretical research.

Of course there are paradigms involving probability which have not been discussed here. These include judgements on natural hazards, e.g. the probabilities of floods, diverse research on forecasting and planning, research on multiple cue probability learning, or the investigation of researchers handling probability in experimental design such as the inadequate consideration of statistical power. Furthermore, there is the growing field of research on risk-assessment and risk-perception including current topics like rating the probability of technical risks like power plants disasters or estimating the spread rate of epidemic diseases like aids.

Overview on research paradigms

Finally, we present a taxonomy of the reviewed research paradigms based on the following categorization: crucial features of the experimental situation, the normative model, the conception of the subject, the dominant psychological explananation, the subject-task relation, and, last but not least, a brief judgement on the didactic value.

Paradigm	Experimental situation	Normative model	Conception of the subject	Psychological modelling	Subject-task relation	Didactic value
Probability learning	physical stimuli (light bulbs etc.)	Bernoulli series, Markov chains	man as probability automaton	learning in probabilistic settings as or other theories linear operator	ontological objectivism	primary intuitions of frequentist probability
Bayesian revision	diagnostic data to judge different hypotheses	Bayes' theorem, likelihood principle	man as Bayesian automaton	gradual revision of opinion by Bayes' theorem,	ontological objectivism	unclear as people do not obey Bayes' rule
Disjunctive & conjunctive probabilities	word problems or urn apparati, response: multiple choice or numerical estimates	addition & multiplication rule		ad hoc explanations and modelling of specific inferences		
Correlation & contingency	2x2 tables, response: multiple choice or numerical estimates	contingency measures	man as theory guided data processor	inferential rules, computational heuristics		insight into erroneous data processing
Judgemental heuristics	word problems, response: multiple choice or numerical estimates	unique probability models with different degree of sophistication	man as systematically biased decision maker	limited set of intuitive judgemental heuristics	ontological cognitivism	cognitive processes of probability judgements
SPT-models	word problems, response: detailed work protocol	probability model including different interpretations of probability	man as goal oriented, flexible but constrained information processor	procedural heuristics and semantic knowledge; models of thought to describe activations of the cognitive system	ontological reconstructivism	stochastic thinking as a specific cognitive activity
Probability calibration	almanac problems, weather forecasts etc.	tests on stimulus vs. response frequencies	man as systematically biased probability automaton	e.g. scale artefacts	distorted ontological objectivism	
Event-related brain potential	physical stimuli	Bernoulli series	e.g. unconscious probability processor, expectancy based information processor	different foundations of subjective probability		

Table 3: Characteristics of the paradigms discussed

3. Critical Dimensions of Educational Relevance

We now consider the question of what can be learnt from the review of paradigms and issues with respect to the teaching of probabilistic ideas. Stochastic thinking is not only a branch of mathematics but also a specific approach of individuals to certain problems. The common feature of the studies cited so far is that individuals are presented with tasks in a laboratory setting. This allows an openness not only in the subject's thinking but also in the investigator's interpretation and construction of normative models.

A teacher can only purposefully construct a successful means for improving the students' ability if (s)he has an awareness of the underlying response behaviour and how it might be mediated. The didactic triangle is a model of the mutual relationship between the task, the learner and the teacher. The essential missing component in psychological research is the relationship between the teacher and the student's task. Hence we must deal with the subject-task relation and how it underlies different research paradigms.

The conception of the task

The conception of the task must be the conception of some individual - the experimenter, the subject, the reader etc. There are at least four types of tasks: - physical stimuli, random apparati, word or letter codified problems, real life problems. We need to consider how such tasks (i) serve to indicate an individual approach to probabilistic structures and, (ii) provide insight into thinking processes as part of learning.

Physical stimuli. Experimental stimuli like light bulbs, acoustic signals or graphically displayed data without a random element being obvious are common. In an experimental setting the subject is usually not informed about the nature of the apparatus nor of the rationale for the study. It is doubtful whether such a task is conceived as probabilistic. Indeed, by trying, for example, to display typical representatives, researchers were prone to commit the fallacies conceptualized by Kahneman and Tversky.

Experimental tasks with random apparati. Typical examples of experimental tasks with random elements are coins, dice, roulette wheels, spinners and urns. These are also commonly used in the classroom. However, such settings play a marginal role in the history of psychological research because of problems of comparability (e.g. of specifications), attainability, relevance and economy.

Word problems. Most psychological probability studies use word problems: either letter codified versions of outcomes of random experiments or problem stories apparently from daily life. The latter type, which includes the flood or the cab problem, can often only be interpreted from the notion of subjective probability. Since there is no random sampling, one cannot rely on objective notions. Clearly, word problems are designed to help analyse the subject's notions of probability. However, a task may be seen very differently if instructions or the situation is slightly altered, so that the instructions as well as the actual problem need to be considered as crucial task variables (Luchins and Luchins, 1950).

Real life problems. Experiments were not run in real life settings, neither in the early research in the behaviouristic paradigm, nor in the recent research in the cognitive paradigm, though probability models may be constructed for them. This is due to the difficulty of investigating cognition and behaviour with problems, which are so unstructured that there is no clear means to classify a successful answer. The extraordinary difficulties in investigating behaviour in real life situations is not only caused by their openness; another feature is the social dynamics of a task which is usually more difficult to control in classroom than in laboratory settings. In reality, tasks are socially negotiated and may not be defined by the individual independently of social constraints.

The conception of the subject

What is actually known about the subject and which abilities and competencies are principally assigned to the individual, essentially differs between different paradigms. Clearly, psychological probability research is a highly specific, and somewhat marginal field within psychology. Nevertheless the same changes in the conception of man or conception of the subject may be identified in probability research as in other branches of psychology. Shulman and Carey (1984) singled out four conceptions of human rationality in cognitive psychology which may easily be identified in psychological probability research; they are man as rational animal, man as irrational animal, man as bounded rational, and man as collectively rational.

Man as a rational animal means

> "to act consistently in its own interest ... or in response to the consistency of its environments and to develop through education and learning, capacities for reason as it matures." (p.501)

According to the associationist and reductionist philosophy of Thorndike (1931)

"learning is connecting. The mind is a man's connection-system." (p.122)
The conception of man as probability automaton of the early research on probability learning may be conceived of as a behavioural reductionist variant of associationism. The rational version of this philosophy may be found in the man as calibrated and well organized but somewhat inert Bayesian automaton. It should be noted that within both approaches probability is conceived of as a general concept and a genuine principle of the human mind which is believed to rule the principles of learning, as learning consists of the development of stimulus-response bonds.

The counterpart of the first perspective is the Freudian conception of man as irrational animal; according to this, man is guided on the one hand by the irrational unconscious and on the other hand by an equally unbalanced and hitherto bumbling irrational superego. The ego was required as a compromise between id and superego. Although the irrational animal conception of Freud plays a rather minor role, this approach in some respect underlies the construct of illusionary correlation and fixed theories as proposed by the Chapmans.

The principles of the bounded rationality suggest that individuals make active use of cognitive strategies and previous knowledge to deal with their memory limitations and their restricted operative and heuristic repertoire. The research of the Chapmans already contains the assumption of man as a theory guided processor of limited cognitive abilities. The conception of man as bounded rational also underlies the research on judgemental heuristics and on the structure and process models of stochastic thinking. The main difference between these research lines is that the former is restricted to a few general heuristics which guide the hitherto 'crippled intuitions' in probability judgements, whereas the latter conceptualizes the structure of qualitatively different analytic and intuitive cognitive strategies.

Man as collectively rational means that human ability and intelligence may only be practised and investigated in the context of social interaction. This conception is not to be found in psychological research. This is somewhat surprising, as much of this research was conducted by social psychologists, who are usually trained to pay attention to contextual aspects of probability judgements. However, in most studies the subjects' ability to use the facilities of his/her environment in order to solve a problem is not investigated, rather the subject's approach is investigated in the 'social vacuum' of a laboratory.

The conception of the subject-task relation

The relationship people have to their environment is crucial; nevertheless this seems to have hardly been reflected in most research described so far. There is an amazingly uniform conception of the subject-task relation across all paradigms and types of tasks; the behaviour of the average subject is analysed according to the standard interpretation of the situation or problem. As Gigerenzer (1987, p.26) points out, this methodology is a strategy to overcome two sorts of subjectivism - of the experimenter and the subject respectively. The desire to meet the ideals of classical science led to the elimination of the experimenter making judgements. This yielded the phenomenon of judged probability without judges, a kind of individual psychology without individuals.

Superficially science is characterized by unequivocal models and theories. Similarly, psychologists did not account for the degrees of freedom in the subject's or experimenter's task interpretation. Often they were unaware of the broad scope of notions of probability as a source of different task interpretations. This is reflected in

> "the assumption that subjective values such as length and subjective probabilities exist independently from the context variables present [in the task or in the instruction]." (Gigerenzer and Murray, 1987, p.67)

Nevertheless, there are variations of subject-task interplay based on objectivism, cognitivism or reconstructivism. Probability learning research starts from the Helmholtzian paradigm of the assumption of unconscious inferences. Cognition and response behaviour consist of reactive, passively acquired stimulus-response bonds with no intervening mental models. This assumes a closed, context free (without random variation) predetermined microcosmic relationship between the individual and his/her laboratory environment. Since ontology can be viewed as the separation of outside reality from an individual's consciousness, this conception of the subject-task relation is labelled as 'ontological-objectivistic'. This also features in other conceptions of man - as an inert Bayesian automaton or as an intuitive statistician within the paradigms of probability parameters and judgemental heuristics.

Judgemental heuristics stress the necessity of starting from the subject's understanding and cognitive representation of the task and differentiates between the information available and the information used; so the subject-task relation is labelled 'ontological cognitivism'. However, early studies did not actually investigate the subject's task representation. They were far from Resnick's optimistic description ...

> "today's assumptions about the nature of learning and thinking are interactionist. We assume that learning occurs as a result of mental constructions of the learner. These constructions respond to information and stimuli in the environment, but they do not mirror them." (Resnick, 1981, p.660).

The SPT-models try to acknowledge the active part of the individual by modelling an evaluative structure and a goal system through empirical research which employs word problems. The models conceptualize the multitude of strategies as intuitive judgemental heuristics and analytic strategies using mathematical knowledge as it is taught in school. This is labelled 'ontological reconstructivism'.

The collectively rational perspective draws on Vygotsky (1978) to view task performance as inference based on two different settings: individual problem solving while working independently and problem solving when working with the cooperation of a teacher as a group. Since the subject-task relation is socially negotiated, this is labelled 'social reconstructivism'. However no probability research has yet been done in psychology in this perspective.

In conclusion, judgements on the relevance of research to other disciplines are often contentious. This is particularly true in evaluating the educational relevance which is highly dependent on individual teachers' perspectives. The crucial point is whether a study provides insight into the cognitive processes linked to learning and teaching probability.

The theoretical value of the Bayesian revision studies is unclear; obviously individuals do not adjust their probabilities according to Bayes' rule. So the conception of man as a rational animal has to be abandoned as one cannot expect any direct impact on mathematics education. The conceptions of man as a bounded rational and of ontological cognitivism govern the judgemental revision studies. Despite its methodological limitations, this branch of research provides seminal insights into the nature of relevant cognitive processes. However, research like the structure and process of thinking, aspires to model inferential processes directly and provides a general framework of the salient cognitive components. This allows for a general structure of thought processes as well as an understanding of an individual's variability in treating a problem (e.g. by postulating different modes of thought). Thus, these models provide meaningful conceptions for understanding and interpreting students' stochastic thinking.

The considerable body of studies investigating behaviour, cognition and psychophysical processes in probability follows different paradigms and conceptions of man and the subject-task relation. Research employs varying methodologies and theoretical backgrounds; from an educational point of view, it is necessary to develop a holistic approach integrating these perspectives.

4. Developmental Approaches on the Acquisition of the Probability Concept

Developmental approaches on the acquisition of probability concepts have hardly been considered in the research reviewed so far. One reason for this is the different focus of interest between behavioural or experimental psychologists and developmental psychologists. Whereas the former are mainly concerned with a description and modelling of the average subject's behaviour, the latter are interested in probability experiments mainly in order to improve their theories of cognitive development.

Piaget and Inhelder (1951) who initiated developmental research on the probability concept predominantly used random situations like urn or spinner tasks. The probability of events is determined either spatio-temporally, or logico-arithmetically. There were no attempts to use situations without symmetry. Nevertheless, unlike some of the research reviewed above, real random apparati were used in developmental studies. Furthermore, the clinical interview method of the Piagetian research is also a distinctive feature.

The second paradigm, probability learning, has already been discussed earlier. Note that the sequences of stimuli in probability learning tasks meet important criteria of random sequences in the objectivist sense. A particularly revealing description of probability learning is given by Messick and Solley (1957).

> "The systematic change in output probability as a function of having more and more experience with an input is called probability learning. Studies in this area have used either adult college students ... or subhuman species such as rats ... or goldfish" (p.23)

The cognitive-developmental approach of Piaget and Inhelder

The most comprehensive research approach dealing with the development of the probability concept is the approach of Piaget and Inhelder (1951). They systematically analyzed the probability concept in children and formulated a theory to explain its development. According to Piaget, cognitive development generally consists of a sequential transformation process during which elementary sensi-motor schemata are restructured into progressively more complex cognitive structures, initially into pre-operative and then into operative ones.

Operative cognitive structures constitute the most important prerequisite for the emergence of logical-mathematical concepts. This is because the growing child is only able to reconstruct the quantitative relations inherent in problems by means of logical operations. This also applies to the child's construction of the probability concept. According to Piaget and Inhelder the very basis for the acquisition of this concept is the ability to distinguish between chance and necessity. The pre-operative child is unable to make this distinction. At this stage of development, the child still lacks the ability to construct logical relations such as cause and effect or similar deductive relations, that are necessary for an understanding of why events occur. In other words, the pre-operative child has not yet developed the cognitive framework on the basis of which he/she can grasp the difference between necessary (and thus predictable) and random (non-predictable) events. Random events, in his/her understanding are still subject to the same deterministic order, which also controls necessary events.

To understand the randomness of events, the child must have organized cognitive structures into concrete operations of thinking. These enable the child to comprehend deterministic events by means of logical relations, and thus to recognize the logical necessity of their occurrence. On the other hand, stochastic events cannot be represented in concrete operative structures because of their irreversibility. Hence, the child cannot infer their occurrence until he/she becomes able to differentiate between necessary and random events.

In a further developmental step beyond this distinction, the child acquires the ability to make chance events operatively calculable. The concrete-operative child may already be able to make some quantitative estimates of probability, but only when the number of events is limited and the proportion between favourable and non favourable cases can be calculated simply. In order to overcome the limitations associated with concrete operations in probability, the child has to construct and represent the totality of all events and to take multiple relations into account, when he/she is calculating proportions. The prerequisites for doing this are acquired in the formal-operative stage, during which the concrete operations are co-ordinated and transformed into a completely closed and reversible system of logical operations.

Fischbein's learning-developmental approach

According to Fischbein (1975), Piaget and Inhelder's stage model of the development of the probability concept has two serious deficiencies. First, the findings on the growth of probability learning cannot be integrated into this model. Fischbein demonstrated that young children may differentiate between certainty and uncertainty. Even three year old children show probability matching (Messick and Solley, 1957), particularly if the tasks do not require a high level of verbal understanding. Secondly, the importance of the learning processes leading to the formation of the probability concept is neglected. As an alternative, Fischbein formulates a learning-developmental approach to the acquisition of the probability concept.

Fischbein postulates that children, even in the pre-operative stage, have a pre-conceptual understanding of both relative frequencies and probabilities which is founded on primary intuitions, a type of implicit knowledge growing spontaneously out of a child's everyday experience. As the child grows older, primary intuitions of relative frequency, chance and probability are successively transformed into an operative concept of probability. This transformation is not the spontaneous and quasi automatic result of a self regulating process of growth, but rather is mediated by instructional intervention at school, for instance by explaining to the child the basic ideas of probability theory and the rules of probability calculus. Without such purposeful intervention, an operative concept of probability cannot be developed, even in adults.

Information processing approaches

Fischbein's *learning*-developmental approach is an alternative to Piaget's *cognitive*-developmental stage model. Information processing provides an even more radical alternative. Three approaches can be identified that are based on heterogeneous theories of cognitive development. Within these approaches the acquisition of competencies is not a function of the development of cognitive structures with respect to their organization into operative systems as is with Piaget. Rather, it depends on both the growth of information processing space and on the changes in information processing and problem solving strategies which occur during the course of development. However, a comprehensive and unitary theory of acquisition of the probability concept has not been elaborated so far. There are some separate studies focusing on the analysis of strategies and abilities in solving probability problems.

Scardamalia (1977) attempted to show that the ability to cope with combinatorial problems is a function of a subject's information processing space and the information processing demands of a task. Combinatorial problems were designed to be invariant with respect to their logical structure but varying in their information processing demands. Subjects with different information processing spaces were asked to work on them. Results showed that even concrete-operative children are ready to solve combinatorial problems provided that the task demands do not exceed their information processing space. Contrary to Piaget and Inhelder, Scardamalia's findings indicate that formal-operative thinking is not a prerequisite for being able to deal with combinatorial problems.

Brainerd (1981) also proposed that the development of probability judgements is adequately explained as a function of the growth of working memory. He designed a general model of the information processing in working memory in solving container tasks; this model contains a total of four storage and three processing operations. In a second step, he modified this model with regard to several variations of the task (for instance with regard to tasks with and without replacement), in such a way that specific effects of each of the four storage and the three processing operations on subject's responses could be predicted and tested by analyzing the frequency of hits and failures observed.

Brainerd conducted a series of experiments with pre-school and primary school children. Their responses indicated that all age groups did well in storing frequency data; however, younger children failed to retrieve the correctly stored data. In other words, the wrong predictions observed were primarily due to retrieval errors. These retrieval errors may not depend on a lack of specific cognitive structures, but rather on limitations of memory space; the decrease of failures in older children is sufficiently explained as a function of their growing memory space.

Finally, Siegler (1986) attempted to prove that information processing rules for solving probability tasks like card or urn experiments change in the course of a child's development. He distinguishes between a dominant dimension of favourable events, and a subordinate dimension of the unfavourable events.

Rule I. The urn with a higher number of favourable events is chosen.

Rule II. With an equal number of favourable and unfavourable events the urn with the smaller number of unfavourable events is selected.

Rule III. The difference between the number of favourable and unfavourable events is calculated for each urn and the one with the greater difference is selected.

Rule IV. The ratio between favourable and unfavourable events is the choice criterion.

As the cognitive development is progressing, these rules are integrated into increasingly complex systems, analogous to binary decision tree models. Which rule is applied in solving a probability problem, depends both on the complexity of a subject's system of rules and on the structure of the task.

In order to test the validity of this assumption, Siegler designed a total of six urn tasks. Favourable and non favourable elements were combined in such a manner that the responses could be directly linked to each of the four rules. These predictions were then compared with the responses of 3 to 20 year old subjects. Subjects' responses agreed very well with the predictions made for Rules I and IV. However, they only partly fitted the response patterns that were expected for Rules II and III. Furthermore, it was quite obvious that pre-school children, with one exception, were applying only Rule I, whereas even the 8 year olds made their probability judgements predominantly according to Rule IV and partly Rule III, skipping II.

Semantic-conceptual and operative knowledge approach

We can now synthesize some earlier propositions. First, probability is not a homogeneous concept but rather a concept field with a set of facets. Second, the individual's concept field (like the scientific concept field) is designed by an extensional and an intensional dimension. The extensional dimension is given by the different facets and by the structure of the relations between the facets, the intensional dimension is constituted by the operative heuristics and the correspondences that are tied to elementary concepts of a specific facet. Third, one and the same situation may be represented by different aspects of probability. Fourth, according to the specific facet, which is elicited in a child, different cognitive processes or modes of thought are involved. Fifth, this view allows for a more differentiated understanding of what the precursor skills of sophisticated stochastic thinking are and what constitutes developmental growth.

These views are exemplified in an experimental study by Scholz and Waschescio (1986); they constructed an experimental design with two spinner-tasks. All tasks could only be solved with the Rule IV above or at the formal stage of Piaget's hierarchy. The subjects had to choose one of the two spinners which was then used in a roulette-like game.

Scholz and Waschescio postulated that subjects would use the equal probability concept as a reference point. They distinguished between different favourability types; a spinner with a winning chance of above 0.5 is called favourable, below 0.5 unfavourable, and equal to 0.5 neutral. Note that one may construct pairs of spinners with the same logico-arithmetical properties, for instance with the same chance difference but belonging to either the same or to different favourability types. Falk et al. (1980) report that tasks with different favourability types are easier than tasks with the same type. The results of Scholz and Waschescio indicate that the equal probability concept serves as a mediating tool. Though not all children use this concept, it is clear that even amongst pre-school children this concept is partly available; tasks with the same favourability type are more difficult than a task comparing a favourable or an unfavourable spinner with a neutral one and these are more difficult than comparing a favourable with an unfavourable one.

Another approach in two spinner tasks is to forget about any counting and numerical representation of the problem and just base one's choice on the holistic visual impression, for instance on the intensity of the winning colour. Perceptual strategies of this preoperative type may be successfully applied in many problems which may be interpreted as geometrical probability. Again some empirical evidence may be found for such a strategy, for instance, in Hoemann and Ross (1981) who called this a magnitude estimation strategy.

Scholz and Waschescio tested this strategy by introducing three levels of chance differences, keeping other salient task characteristics (including the favourability type) constant. Thus, for instance the chance difference in a pair with 6 reds and 7 blues vs. 2 reds and 3 blues is 0.06 while the chance difference in a pair with 5 reds and 8 blues vs. 2 reds and 5 blues is 0.01. The data suggest that perceptual strategies are available even in pre-school children. Clearly, within the terminology of the structure and process of thinking approach, the favourability inference is a typical analytic strategy, or a conscious subjective probability; on the other hand the perceptual auxiliary strategies are intuitive ones, they may be applied in situations that are interpreted or represented in terms of geometrical probability. Thus the type of inference and the type of operations applied depends on the semantic structure and representation of a situation.

Discussion of the developmental approaches

The conception and small range of tasks in developmental research were mentioned in the introduction. Developmental research has primarily focused on symmetrical chance experiments. Though other facets of probability like logical or subjective probability were identified by various psychologists (e.g. Cohen and Hansel, 1955; Konold, 1983), a systematic theory of development is still missing.

Let us briefly examine which conception of the child and its development is established by the different approaches. To begin with, one should notice that a pure ontological position is not covered by any of the reviewed approaches. Developmental psychologists who conduct research on the probability concept are usually aware of the necessity of representing and reconstructing a task. Consequently, their conception of the subject-task relation is either of an ontological cognitivism or an ontological reconstructivism.

Piaget and Inhelder presume that operative-cognitive structures develop and subjects do not successfully solve probability tasks till logical and combinatorial thinking is at its highest stage. In their framework, ontological reconstructivism means analytic, logical, numerical-conceptual representation and problem solving. This is a one sided view of cognitive development as shown in various ways.

Information processing theories revealed that cognitive development may not be restricted to the development of operative heuristics. Case (1977) and Pascual-Leone (1978) have demonstrated that development is accompanied by an age specific extension of the memory space. Furthermore, what is known or possibly available has to be retrieved appropriately from memory.

One may presume that the specific notion of probability attributed is quasi automatically determined by the specific memory load and that this process is established by simple retrieval relations. However, this assumption would provide a rather limited conception of the individual's activity in interpreting and reconstructing a situation. Whether a qualitative view is taken referring to verbal labels like 'probable', 'not likely' etc. or a quantitative approach is preferred may also be the object of an individual's decision process. Within the framework of the structure and process models of thinking, these comparative 'decisions' are modelled by evaluative operations.

Finally, the development of probability is the development of intuitive abilities and of analytic thinking. Though starting from a somewhat different definition of intuitive thinking than that provided here, Fischbein (1987) emphasizes the role of intuitions in the course of development and learning. Yet there is still no general cognitive or developmental-cognitive theory on the acquisition of probability.

Bibliography

Arkes, H.R. and A.R. Harkness: 1983, 'Estimates of Contingencies between two Dichotomous Variables', *Journal of Experimental Psychology, General* **112**(1), 117-135.

Bar-Hillel, M.: 1980, 'The Base-Rate Fallacy in Probability Judgments', *Acta Psychologica* **44**, 211-233.

Birnbaum, M.H.: 1983, 'Base-Rates in Bayesian Inference: Signal Detection Analysis of the Cab Problem', *American Journal of Psychology* **96**(1), 85-94.

Brainerd, Ch.J.: 1981, 'Working Memory and the Developmental Analysis of Probability', *Psychological Review* **88**, 463-502.

Brunswik, E.: 1955, 'Representative Design and Probabilistic Theory in a Functional Psychology', *Psychological Review* **62**, 193-217.

Brunswik, E. and H. Herma: 1951, 'Probability Learning of Perceptual Cues in the Establishment of a Weight Illusion', *Journal of Experimental Psychology* **41**, 281-290.

Bush, R.R. and F. Mosteller: 1955, *Stochastic Models for Learning*, Wiley, New York.

Case, R.: 1978, 'Intellectual Development from Birth to Adolescence: A Neo-Piagetian Interpretation', in R.S. Siegler (ed.): *Children's Thinking: What Develops?*, Lawrence Erlbaum, Hillsdale, 37-71.

Chapman, L.J. and J.P. Chapman: 1967, 'Genesis of Popular but Erroneous Diagnostic Observations', *Journal of Abnormal Psychology* **72**, 193-204.

Cohen, J., E.J. Dearnaley and C.E.M. Hansel: 1956, 'The Addition of Subjective Probabilities', *Acta Psychologica* **12**, 371-380.

Cohen, J. and C.E.M. Hansel: 1955, 'The Idea of Independence', *British Journal of Psychology* **46**, 178-150.

Donchin, E. and M.G.H. Coles: 1988, 'Is the P300 Component a Manifestation of Context Updating?', *Behavioral and Brain Sciences*, **11** (3), 343-356.

Edwards, W.: 1968, 'Conservatism in Human Information Processing', in B. Kleinmuntz (ed.): *Formal Representation of Human Judgment*, Wiley, New York, 17-52.

Estes, W.K.: 1964, 'Probability Learning', in A.W. Melton (ed.): *Categories of Human Learning*, Academic Press, New York, 89-128.

Falk, R., R. Falk, and I. Levin: 1980, 'A Potential for Learning Probability in Young Children', *Educational Studies in Mathematics* **11**, 181-204.

Falk, R. and D. MacGregor: 1983, 'The Surprisingness of Coincidences', in P. Humphreys, O. Svenson, and A. Vari (eds.): *Analysing and Aiding Decision Processes*, North Holland, Amsterdam, 489-502.

Fischbein, E.: 1975, *The Intuitive Sources of Probabilistic Thinking in Children*, Reidel, Dordrecht.

Fischbein, E.: 1976, 'Probabilistic Thinking in Children and Adolescents', in *Materialien und Studien vol.2, Forschung zum Prozeß des Mathematiklernerns*, IDM: Bielefeld, 23-42.

Fischbein, E.: 1987, *Intuition in Science and Mathematics. An Educational Approach*, Reidel, Dordrecht.

Flood, M.M.: 1954, 'Environmental Non-Stationarity in a Sequential Decision Making Experiment', in R.M. Thrall, C.H., Coombs, and R.L. Davis (eds.): *Decision Processes*, Wiley, New York, 287-299.

Gigerenzer, G.: 1987, 'Probabilistic Thinking and the Fight against Subjectivity', in L. Krüger, G. Gigerenzer, and M.S. Morgan (eds.): *The Probabilistic Revolution, vol.II: Ideas in the Sciences*, MIT Press, Cambridge, MA, 11-33.

Gigerenzer, G. and D.J. Murray: 1987, *Cognition as Intuitive Statistics*, Lawrence Erlbaum, Hillsdale N.J.

Goulet, L.R. and K.S. Goodwin: 1970, 'Development and Choice Behavior in Probabilistic and Problem-solving Tasks', in H.W. Reese and L.P. Lippsitt (eds.): *Advances in Child Development and Behavior*, Academic Press, New York, 213-254.

Green, D.M. and J.A. Swets: 1966, *Signal Detection Theory and Psychophysics*, Wiley, New York.

Helson, H.: 1964, *Adaption Level Theory*, Wiley, New York.

Hoemann, H.W. and B.M. Ross: 1981, 'Children's Concept of Chance and Probability Concepts', *Child Development*, **42**, 221-236.

Hogarth, R.M.: 1980, *Judgement and Choice: The Psychology of Decision*, Wiley, New York.

Horst, R.L., R.J. Johnson, and E. Donchin: 1980, 'Event Related Brain Potentials and Subjective Probability in a Learning Task', *Memory and Cognition* **8**, 476-488.

Humphreys, P.C. and A.D. Wishuda: 1987, *Methods and Tools for Structuring and Analysing Decision Problems, vol.1: A Review, vol.2: A Catalogue*, Technical Report 87 - 1, London School of Economics and Political Science, London.

Jarvik, M.E.: 1951, 'Probability Learning and a Negative Recency Effect in the Serial Anticipation of Alternative Symbols', *Journal of Experimental Psychology* **41**, 291-297.

Kahneman, D. and A. Tversky: 1972, 'Subjective Probability: A Judgment of Representativeness', *Cognitive Psychology* **3**, 430-454.

Kahneman, D. and A. Tversky: 1973, 'On the Psychology of Prediction', *Psychological Review* **80**(4), 237-251.

Kahneman, D. and A. Tversky: 1982, 'Variants of Uncertainty', in D. Kahneman, P. Slovic and A. Tversky (eds.): *Judgment under Uncertainty, Heuristics and Biases*, Cambridge University Press, Cambridge, 509-520.

Kahneman, D., P. Slovic, and A. Tversky: 1982, *Judgment under Uncertainty, Heuristics and Biases*, Cambridge University Press, Cambridge.

Karis, D., G.L. Chesney, and E. Donchin: 1983, ' "... 'twas Ten to One; and Yet We Ventured ...": P300 and Decision Making', *Psychophysiology*, **20**, 260-268.

Konold, C.: 1983, *Conceptions of Probability: Reality between a Rock and a Hard Place*, Unpublished Ph.D. Thesis, University of Mass, Amherst.

Lee, W.: 1971, *Decision Theory and Human Behavior*, Wiley, New York.

Lichtenstein, S., B. Fischhoff, and D. Phillips: 1982, 'Calibration of Probabilities: The State of the Art to 1980', in D. Kahneman, P. Slovic, and A. Tversky (eds.): *Judgment under Uncertainty: Heuristics and Biases*, Cambridge University Press, Cambridge, 306-334.

Luchins, A.S. and E.H. Luchins: 1950, 'New Experimental Attempts at Preventing Mechanization in Problem Solving', *Journal of General Psychology* **42**, 279-297.

May, R.S.: 1986, 'Overconfidence as a Result of Incomplete and Wrong Knowledge', in R.W. Scholz (ed.): *Current Issues in West German Decision Research*, 13-30.

Messick, S.J. and C.M. Solley: 1957, 'Probability Learning in Children: Some Exploratory Studies', *Journal of Genetic Psychology* **90**, 23-32.

Murphy, A.H. and R.L. Winkler: 1974, 'Probability Forecasts: A survey of National Weather Service Forecasts', *Bulletin of the American Meteorological Society* **55**, 1449-53.

Nisbett, R.E. and L. Ross: 1980, *Human Inference: Strategies and Short-comings of Social Judgment*, Prentice-Hall, Englewood Cliffs, N.J.

Pascual-Leone, J.: 1978, 'On Learning and Development, Piagetian Style I and II', *Canadian Psychological Review*, 270-297.

Piaget, J. and B. Inhelder: 1951/1975, *La Genèse de l'Idée de Hasard chez l'Enfant*, Presses Universitaires de France, Paris, translated as: *The Origin of the Idea of Chance in Children*, Norton, New York.

Pitz, G.F.: 1974, 'Subjective Probability Distributions for Imperfectly Known Quantities', in L.W. Gregg (ed.): *Knowledge and Cognition*, Wiley, New York, 29-41.

Rapoport, A. and T.S. Wallsten: 1972, 'Individual Decision Behavior', *Annual Review of Psychology*, 131-176.

Resnick, L.B.: 1981, 'Instructional Psychology', *Annual Review of Psychology* **32**, 659--704.

Restle, F.: 1966, 'Run Structure and Probability Learning: Disproof of Restle's Model', *Journal of Experimental Psychology* **72**, 751-760.

Ross, B.M. and J.F. de Groot: 1982, 'How Adolescents Combine Probabilities', *The Journal of Psychology* **110**, 75-90.

Scardamalia, M.: 1977, 'Information Processing Capacity and the Problem of Horizontal Decalage: A Demonstration Using Combinatorical Reasoning Task', *Child Development* **48**, 28-37.

Scholz, R.W.: 1987, *Cognitive Strategies in Stochastic Thinking*, Reidel, Dordrecht.

Scholz, R.W. and M.B. Köntopp: 1990, 'Elemente der heuristischen Struktur bei Wahrscheinlichkeitsurteilen', in K. Haussmann and M. Reiss (eds.): *Mathematische Lehr-, Lern-, Denk-Prozesse*, Hogrefe, Göttingen, 107-130.

Scholz, R.W. and R. Waschescio: 1986, 'Childrens' Cognitive Strategies in Two-Spinner Roulette Tasks', in *Proceedings of the Tenth Intern. Conf. Psychology of Mathematics Education*, University of London, London, 463-468.

Shulman, L.S. and N.B. Carey: 1984, 'Psychology and the Limitations of Individual Rationality, Implications for the Study of Reasoning and Civilty', *Review of Educational Research* **54** (4), 501-524.

Siegler, R.S.: 1986, *Children's Thinking*, Prentice Hall, Englewood Cliffs.

Slovic, P., S. Lichtenstein, and B. Fischhoff: 1986, 'Decision Making', in R.C. Atkinson, R.J. Herrnstein, G. Lindzey, and R.D. Luce (eds.): *Steven's Handbook of Experimental Psychology*, 2nd ed., Wiley, New York, 673-738.

Svenson, O.: 1981, 'Are We All Less Risky and More Skillful than Our Fellow Drivers?', *Acta Psychologica* **47**, 143-148.

Thorndike, E.L.: 1931, *Human Learning*, Century, New York.

Tversky, A. and D. Kahneman: 1973, 'Availability: A heuristic for Judging Frequency and Probability', *Cognitive Psychology* **3**, 207-232.

Tversky, A. and D. Kahneman: 1979, 'Causal Schemas in Judgments under Uncertainty', in M. Fishbein (ed.), *Progress in Social Psychology*, Lawrence Erlbaum, Hillsdale, 49-72.

Tversky, A. and D. Kahneman: 1983, 'Extensional Versus Intuitive Reasoning: The Conjunction Fallacy in Probability Judgment', *Psychological Review* **90** (4), 293-315.

Verleger, R.: 1988, Event-related Potentials and Memory: A Critique of the Context Updating Hypothesis and an Alternative Interpretation of P300, *The Behavioral and the Brain Sciences* **3**, 343-356.

Vygotsky, L.S.: 1978, *Mind in Society: The Development of Higher Psychological Processes*, Cambridge University Press, Cambridge.

Winterfeldt, D. v. and W. Edwards: 1986, *Decision Analysis and Behavioral Research*, Cambridge University Press, Cambridge.

Roland W. Scholz
Institut für Didaktik der Mathematik
Universität Bielefeld
Postfach 8640
DW-4800 Bielefeld
FR. Germany

Looking Forward

In a sense, this book has raised more questions than it has answered; of course it is the process that is important rather than glibly presented solutions. Nevertheless, we hope that a few broad themes have emerged. For probability is certain to remain as an important topic in school mathematics. Less than four decades ago, it only featured in higher mathematics; it is remarkable that aspects of probability are now even taught to infants.

We have shown the long and sometimes tortuous emergence of probability historically. Gambling was a popular pastime for several millennia but appropriate mathematical models have only been developed relatively recently. The importance of remembering the theoretical nature of stochastic knowledge has been clearly brought out. Curricular initiatives have not always respected this and have tended to be over-ambitious; we have exemplified the means whereby they can be judged. In particular, ideas for use in classroom are brought out in the discussion of empirical research, whilst psychological constraints also need to be borne in mind. The availability of computers is having an increased effect, particularly in the more advanced countries.

In each of the chapters there are suggestions for further research: developing different items to illuminate children's concepts of probability, investigating the extent to which a planned curriculum actually becomes one that students exercise later, the development of a wider range of exemplary task systems, the effectiveness of computers in certain situations, constructing a cognitive theory on the acquisition of probability, etc. But we will not attempt to draw up a comprehensive list here, rather the reader can select appropriate questions within the context of individual chapters.

In more general terms for future research, there is certainly a need for a comprehensive book on the teaching of statistics. There are many individual books dealing with specific aspects, particularly offering suggestions for classroom approaches. But there is, as yet, no global research text offering an international perspective. This could also explore the links between probability and statistics, a controversy we have deliberately avoided. It is a debate which continues within the academic discipline in itself so that a systematic educational analysis is now overdue given the experience of children being taught these ideas in school for 20 years or more.

Finally we end with a few provocative statements which have been raised in various parts of the book as ideas to build up a theory of probabilistic development.

(1) People use personal experience in assessing chance in a haphazard manner.

(2) People process information in a rather incomplete way.

(3) People process information in a way biased by memorable events.

(4) People find it hard to assess probabilities which are very low or very high.

(5) People do not assign values of 0 for impossibility and 1 for certainty.

(6) People equate certainty and impossibility with physical rather than logical events.

(7) People equate 50-50 chances with coin tossing.

(8) People assign equal likelihood in unknown situations.

(9) People are incoherent in assigning and in processing probabilities.

(10) People are supra-additive.

Reference

Wright, G. and P. Whalley: 1983, 'The Supra-additivity of Subjective Probability', in B. Stigum (ed.): *Foundations of Utility and Risk Theory with Applications*, Reidel, Dordrecht.

Index

Mathematics Education Library

Managing Editor: A.J. Bishop, Cambridge, U.K.

1. H. Freudenthal: *Didactical Phenomenology of Mathematical Structures*. 1983 ISBN 90-277-1535-1; Pb 90-277-2261-7
2. B. Christiansen, A. G. Howson and M. Otte (eds.): *Perspectives on Mathematics Education*. Papers submitted by Members of the Bacomet Group. 1986. ISBN 90-277-1929-2; Pb 90-277-2118-1
3. A. Treffers: *Three Dimensions*. A Model of Goal and Theory Description in Mathematics Instruction – The Wiskobas Project. 1987 ISBN 90-277-2165-3
4. S. Mellin-Olsen: *The Politics of Mathematics Education*. 1987 ISBN 90-277-2350-8
5. E. Fischbein: *Intuition in Science and Mathematics*. An Educational Approach. 1987 ISBN 90-277-2506-3
6. A.J. Bishop: *Mathematical Enculturation*. A Cultural Perspective on Mathematics Education. 1988 ISBN 90-277-2646-9; Pb (1991) 0-7923-1270-8
7. E. von Glasersfeld (ed.): *Radical Constructivism in Mathematics Education*. 1991 ISBN 0-7923-1257-0
8. L. Streefland: *Fractions in Realistic Mathematics Education*. A Paradigm of Developmental Research. 1991 ISBN 0-7923-1282-1
9. H. Freudenthal: *Revisiting Mathematics Education*. China Lectures. 1991 ISBN 0-7923-1299-6
10. A.J. Bishop, S. Mellin-Olsen and J. van Dormolen (eds.): *Mathematical Knowledge: Its Growth Through Teaching*. 1991 ISBN 0-7923-1344-5
11. D. Tall (ed.): *Advanced Mathematical Thinking*. 1991 ISBN 0-7923-1456-5
12. R. Kapadia and M. Borovcnik (eds.): *Chance Encounters: Probability in Education*. 1991 ISBN 0-7923-1474-3

KLUWER ACADEMIC PUBLISHERS – DORDRECHT / BOSTON / LONDON

9789401055635